Particle Tracking

Computational Strategies and Diverse Examples

by D. James Benton

Preface

Tracking particles through a moving field, such as flowing air, ground, or surface water, is not only a powerful computational method but can also provide insightful visualizations. I have employed this process for a variety of problems from airborne contamination, thermal pollution, and contaminant transport in groundwater. I have even used it to successfully model the mixing of fluorescent dye through the confluence of three pumps in order to distinguish the time of arrival at the destination. While the Lagrangian method (based on time as the independent parameter) is most often used, the Hamiltonian method (based on time as a dependent parameter) can be much faster. Both are developed in this text, which includes many examples. Diffusion (spreading based on material properties) and also dispersion (based on local velocities) are derived, implemented, and illustrated in the text. Flow in factures within porous media, essential to accurately modeling karst formations, is also covered.

All of the examples contained in this book,
(as well as a lot of free programs) are available at...
https://www.dudleybenton.altervista.org/software/index.html

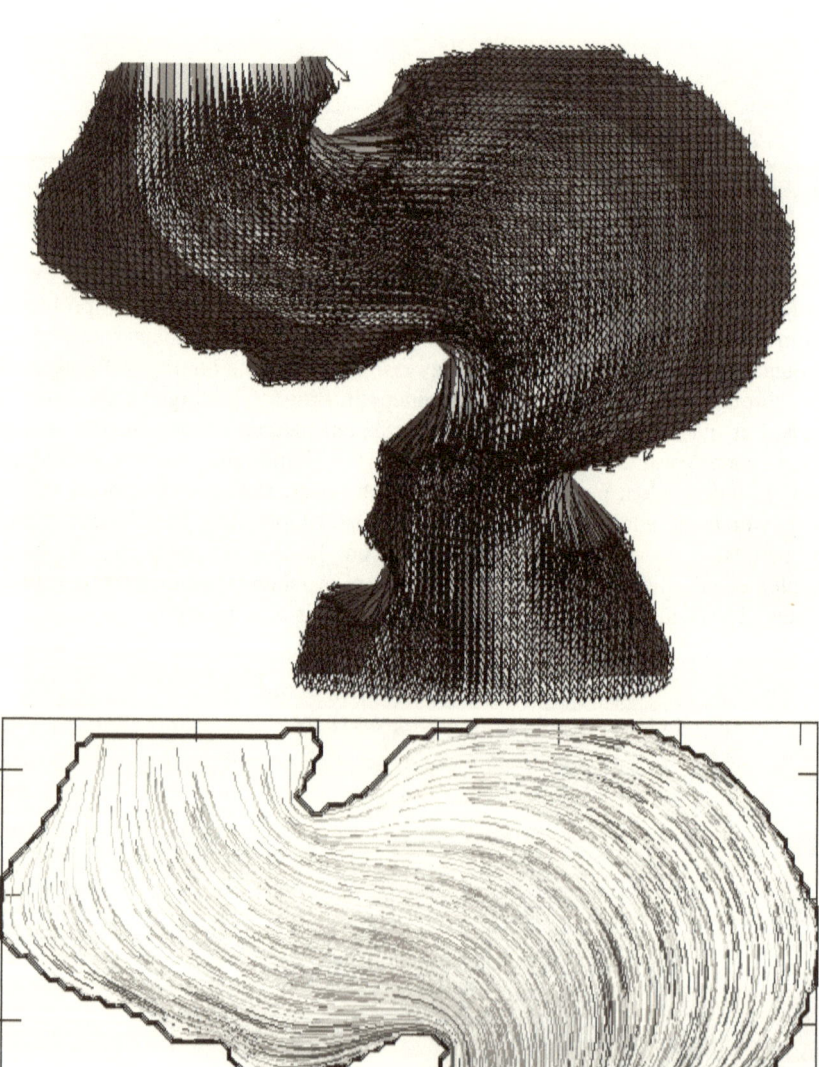

Table of Contents page

Preface i
Chapter 1. Introduction 1
Chapter 2. Two-Dimensional Lagrangian Tracking 3
Chapter 3. Three-Dimensional Lagrangian Tracking 9
Chapter 4. Particle Tracking in Discrete Domains...................................... 11
Chapter 5. Hamiltonian Particle Tracking 17
Chapter 6. Diffusion and Dispersion... 29
Chapter 7. Flow in Fractures.. 45
Chapter 8. Contaminant Plumes... 49
Chapter 9. Particle Seeds 53
Chapter 10. Animations 61
Chapter 11. Concentration Mappings 65
Chapter 12. Reverse Particle Tracking... 67
Chapter 13. Sources and Sinks... 69
Chapter 14. Mosquito Tracking 73
Chapter 15. Tracking Particles Inside Pipes 77
Appendix A. Potential Fields... 83
Appendix B. Boundary Element Method.. 85
Appendix C. Explicit Runge-Kutta Methods 87
Appendix D. Validation of PTRAX... 91
Appendix E. PTRAX Coding 97

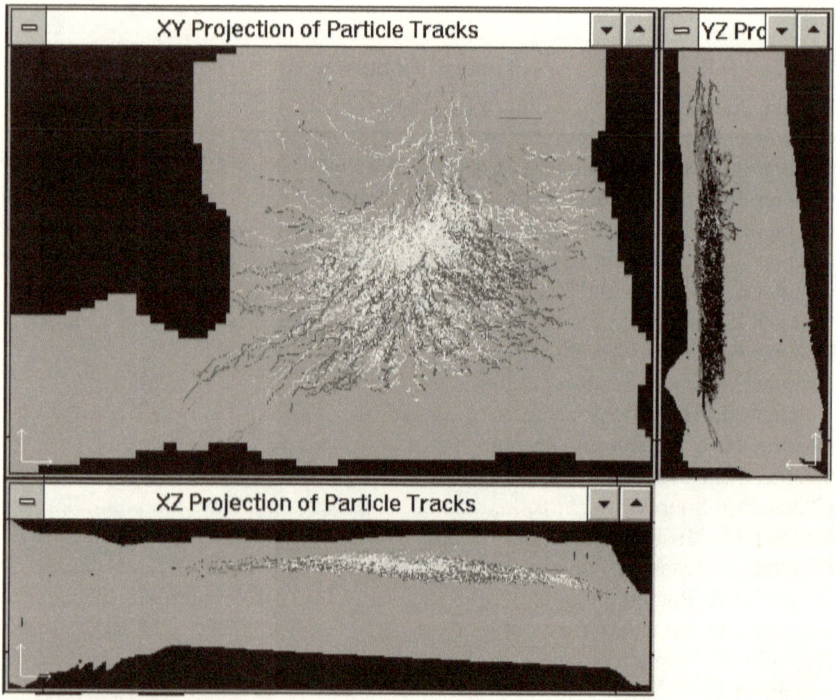

XY Projection of Particle Tracks

YZ Pro

XZ Projection of Particle Tracks

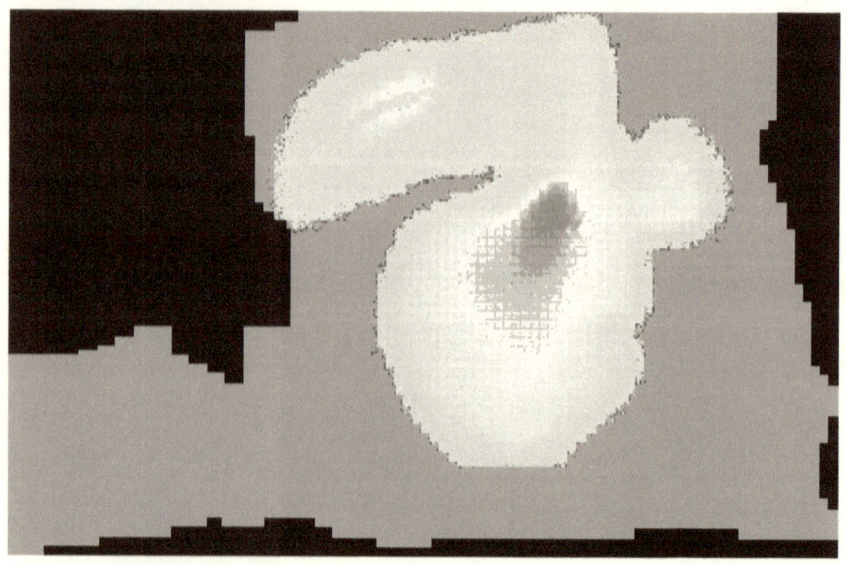

Chapter 1. Introduction

This text focuses on tracking particles within a stationary domain, which is characterized by localized material properties and velocity vectors. The domains are defined by nodes and elements. Nodes define the corners of the elements, not at the center. Velocities are defined at the center of each element. Nodes and elements can be two- or three-dimensional. Two-dimensional elements include: triangles and quadrangles (i.e., four-sided polygons with no particular constraint, such as square corners or parallel sides). Three-dimensional elements include: tetrahedra and bricks (i.e., six-faced polyhedra with no particular constraint, such as square corners or parallel faces). Outward normal faces are assumed, but are not necessary, as software can easily swap vertices to obtain the preferred orientation. It is assumed that the reader already understands these terms.

Eulerian Point-of-View

How the flow fields are obtained (i.e., analytical solution, finite difference method, finite element method, boundary element method, etc.) is immaterial and will only be discussed in passing (see Appendix B for the boundary element method). Suffice it to say, you will need to generate these somehow, which is the subject of another text. The domains and velocity fields so described are from the Eulerian[1] point-of-view, which basically means that the domain is fixed and forms the frame of reference, while flow passes through the domain, varying with position and perhaps time. An example of this perspective would be standing on a bridge watching the river flow and boats float by.

Lagrangian Point-of-View

The Lagrangian[2] point-of-view is not stationary, but moves through the domain, focused on some particular moving item, such as a particle. An example of this perspective would be a passenger on one of the boats, noting the bridge and first observer as the two move relative to each other. This is the classical approach to particle tracking. Tracking a particle using this methodology answers the question: how far will the particle move (and in what direction) over the course of some time step?

Hamiltonian Point-of-View

The Hamiltonian[3] point-of-view is far less common and not so easily illustrated. If you search the Web for descriptions of this approach (e.g., Wikipedia), you will find some rather vague descriptions. Peruse enough of these and a pattern will emerge: the dependent variables discussed are neither space nor time, but quantities like momentum and energy. When it comes to particle tracking, the independent variable is velocity or displacement in the case

---

[1] Leonhard Euler (1707–1783) Swiss mathematician, physicist, astronomer, geographer, logician, and engineer.
[2] Joseph-Louis Lagrange (1736–1813) Italian mathematician and astronomer.
[3] William Rowan Hamilton (1805–1865) Irish mathematician and physicist.

of diffusion or dispersion. Time is a dependent variable. Tracking a particle using this methodology answers the question: how long will it take the particle to move this far? In either case (Lagrangian or Hamiltonian) we must be concerned with the particles leaving one element and entering another, as this is a discrete and not necessarily continuous domain, although we will begin with simple continuous domains and analytical flow solutions.

Chapter 2. Two-Dimensional Lagrangian Tracking

In order to illustrate the process, we need a two-dimensional analytical velocity field. Potential flow theory is the logical choice here. We will not explain potential flow here only utilize it. There is some discussion in Appendix A. Should you need an explanation and derivation, there are numerous sources available on the Web. The code (potflow.c) we will use in this chapter comes from my book, *Complex Variables*, and can be found in the online archive in folder examples\potflow\2D. We begin with one of the simplest fields:

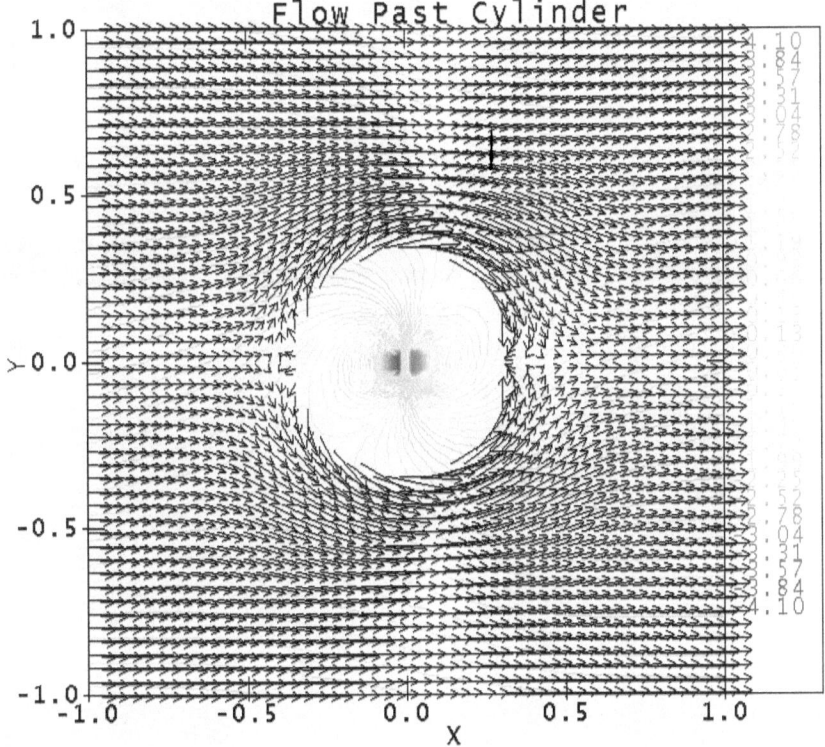

The X and Y components of the velocity vectors (little arrows with tail length proportional to magnitude) are readily calculated at any location, as are the streamlines (colored curves). We will use the velocities, but not the streamlines. Of course, the particles should approximately follow the streamlines, else we've done something wrong. The equations of motion are trivial and arise directly from the definition of the velocity components.

$$u = \frac{\partial x}{\partial t} \quad v = \frac{\partial y}{\partial t} \tag{2.1}$$

3

$$x = x_0 + \int_0^t u\,dt$$
$$y = y_0 + \int_0^t v\,dt \tag{2.2}$$

The initial position (x_0, y_0) is where we drop the particle at time, t=0. The velocity components (u and v) are functions of the position (x,y). We could use any number of integration techniques, such as Runge-Kutta, the simplest being fully-explicit forward Euler:

$$x_{t+\Delta t} = x_t + u\Delta t$$
$$y_{t+\Delta t} = y_t + v\Delta t \tag{2.3}$$

We simply drop a few particles along the boundary at X=-1 and step through time. The code (track1.c) contains the necessary parts of the first (potflow.c) and allows you to easily select which flow field: flow over a cylinder, flow over a cylinder with circulation, flow past a doublet, flow past a half-body, flow past a Rankine body (an ellipsoid), flow past a source, flow past a sink, flow past a source and sink, stagnation flow (against a wall), and uniform (horizontal) flow. The result should be no surprise:

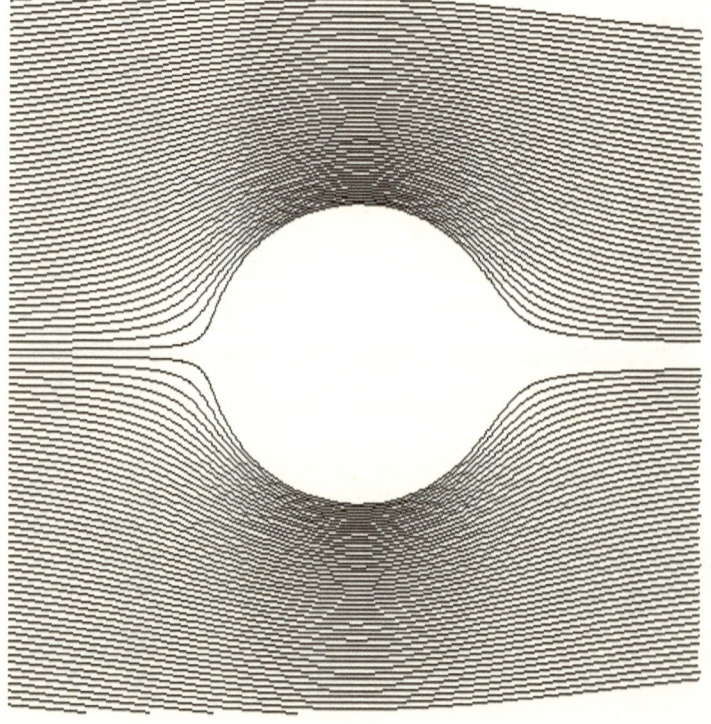

The code couldn't be any simpler:

```
for(n=0;n<seeds;n++)
  {
  x=-1.;
  y=-1.+(n+1)*2./(seeds+1);
  t=0.;
  fprintf(fp,"%lG %lG %lG\n",x,y,t);
  for(i=0;i<max_steps;i++)
    {
    p=flow(x,y);
    x+=time_step*p.u;
    y+=time_step*p.v;
    t+=time_step;
    fprintf(fp,"%lG %lG %lG\n",x,y,t);
    if(x>=1.||y<-1.||y>1.)
      break;
    if(hypot(p.u,p.v)<FLT_EPSILON)
      break;
    }
  }
```

We turn on circulation and obtain:

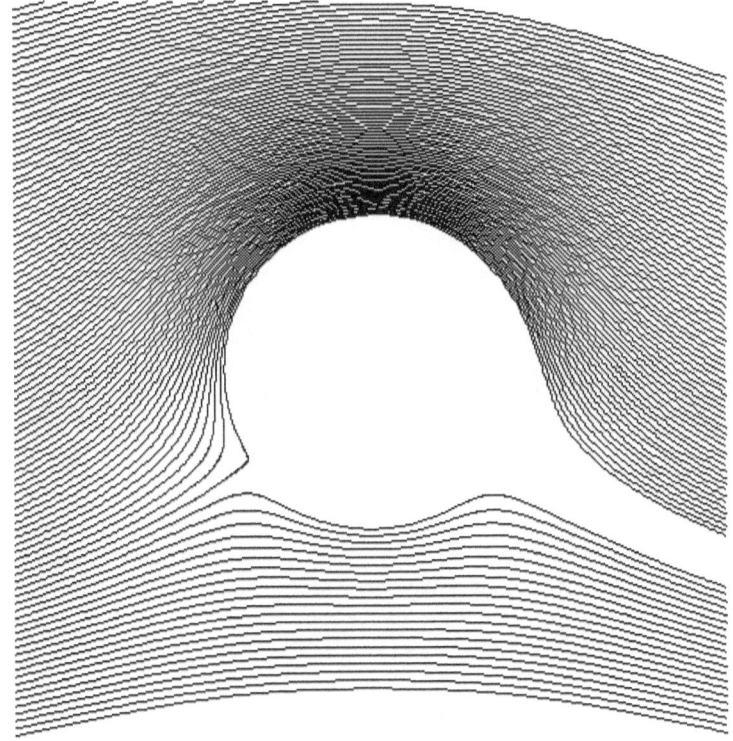

For a flow stagnating at the horizontal plane passing through the middle of the domain, we sprinkle particles along the top and bottom boundaries to obtain:

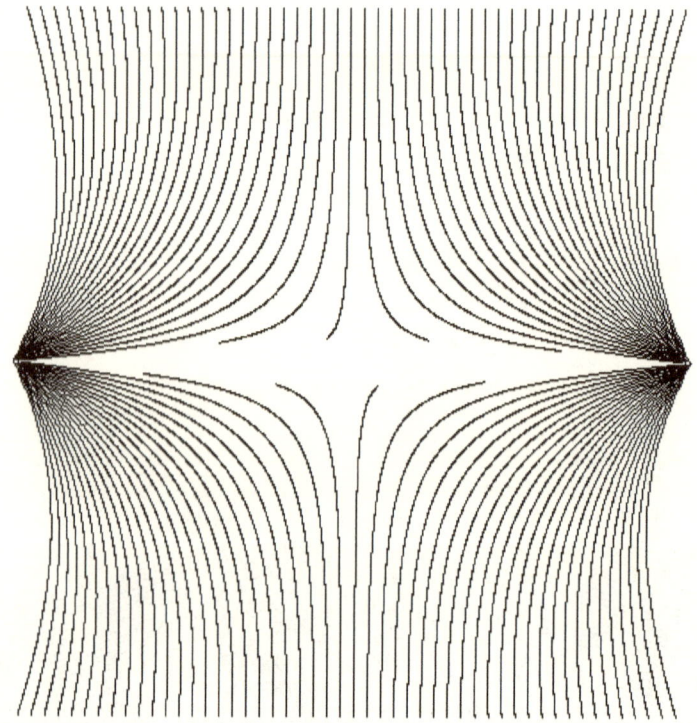

While there's no need here, we will implement a 4th-order Runge-Kutta procedure. I cover the entire family of such methods in my book, *Differential Equations*, showing by example that there is no real advantage to higher orders or implicit variants or error estimates or step-length control embellishments beyond the basic method. While these additions are of historical and theoretical interest, they are of little practical value, as their effectiveness is entirely unpredictable. Simply decreasing the time step until the results level off will always work if anything will. That is, if this doesn't work, then don't expect anything more elaborate to work either. More details on the Runge-Kutta method can be found in Appendix C.

The main advantage of using Runge-Kutta instead of the forward Euler method is greater accuracy with a larger time step, which may or may not be computationally as efficient as using one-fourth the time step, which would still take less processing time than 4th-order Runge-Kutta. Still, I have selected the parameters to illustrate a non-trivial example. The same particles flowing over a cylinder using 4th-order Runge-Kutta are shown in this next figure. The code (track2.c) can be found in the same folder.

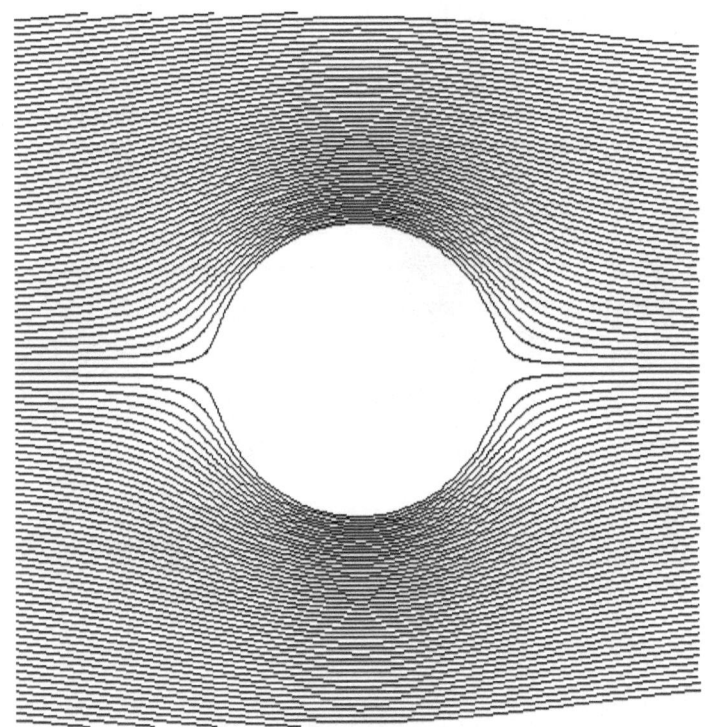

The differences can be seen on the on the downstream side (back) of the cylinder.

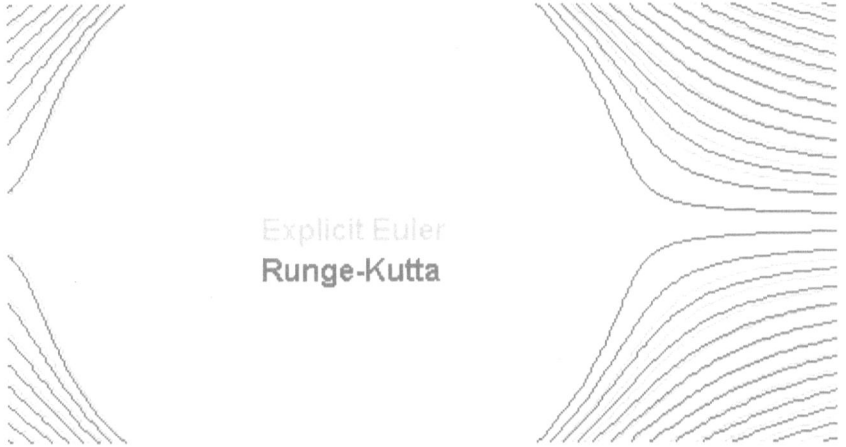

It is important to note that the velocity components and streamlines are symmetric, as this is potential (i.e., inviscid) flow. The velocity components don't change more abruptly on the back side than the front. That's not what's happening here. What causes these two sets of particle tracks to diverge is the effects of small cumulative differences in the calculations. Even a small difference in trajectory early in the history of a particle can result in a large difference in where it ends up, much like the metaphor of a ship being slightly off course on a long journey. See how tightly the Runge-Kutta handles the circulation case:

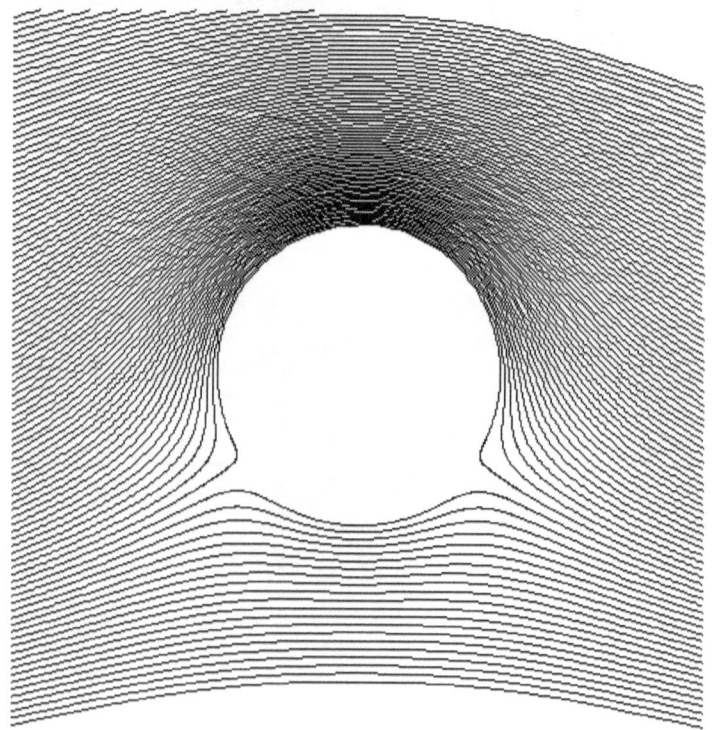

Chapter 3. Three-Dimensional Lagrangian Tracking

The transition from two- to three-dimensional particle tracking is conceptually simple. The biggest implementation difference is in graphically displaying the results on a flat page. The additional formulas are:

$$z = z_0 + \int_0^t w\,dt \qquad (3.1)$$

$$z_{t+\Delta t} = z_t + w\Delta t \qquad (3.2)$$

There are fewer analytical solutions for 3D potential flow than for 2D. The flow code (potflow.c) is in the online archive in folder examples\potflow\3D. The explicit Euler (track1.c) and Runge-Kutta Lagrangian tracking codes are also in this folder. The particle tracks over a sphere are:

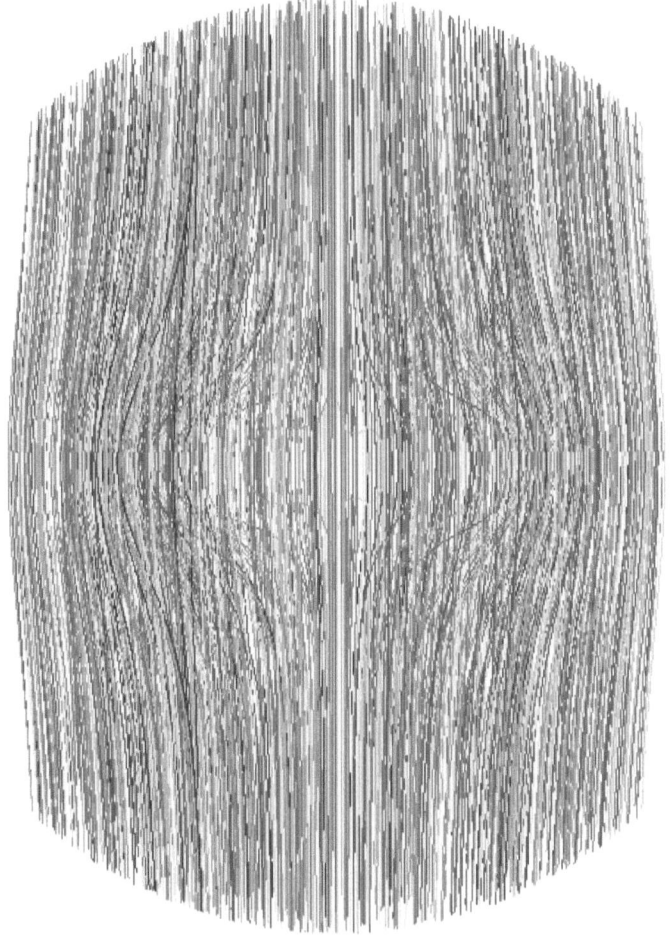

The Runge-Kutta implementation (track2.c) produces tighter particle tracks, as for the 2D examples. This next figure focuses on the zone with the most curvature in the particle paths and also fewer particles.

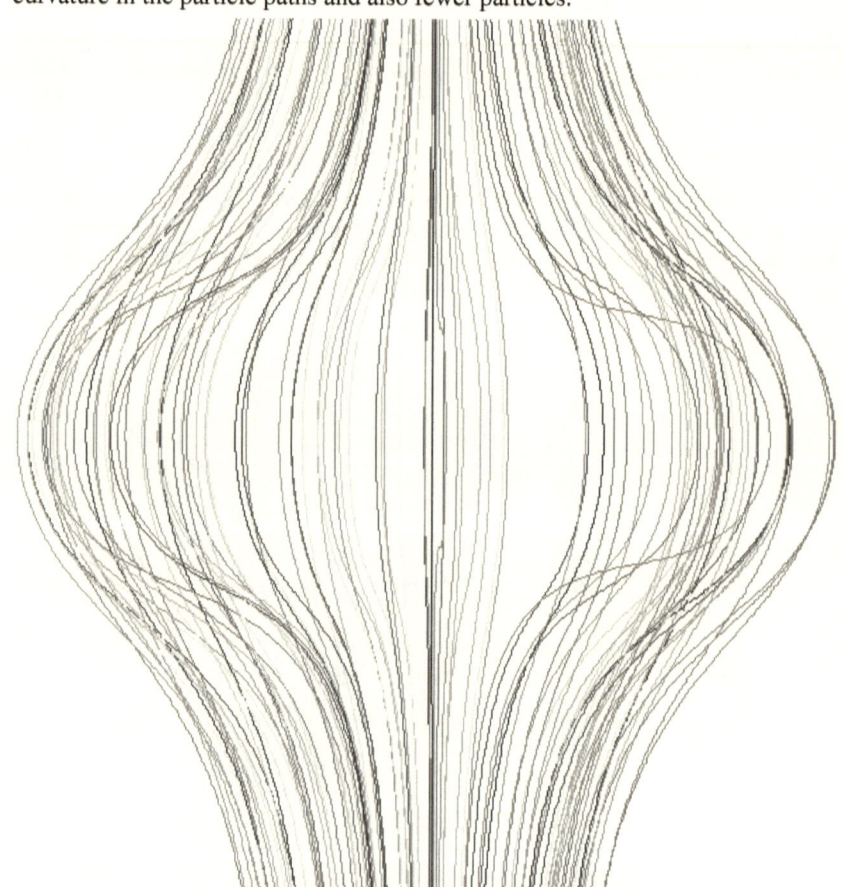

Implementation of Runge-Kutta is very simple:

```
for(i=0;i<max_steps;i++)
  {
  p=flow(y[0],y[1],y[2]);
  if(hypot3(p.u,p.v,p.w)<FLT_EPSILON)
    break;
  RungKutta4(step,&t,time_step,y,dy,3);
  r=hypot3(y[0],y[1],y[2]);
  if(r>3.*D)
    break;
  fprintf(fp,"%1G %1G %1G %1G\n",y[0],y[1],y[2],t);
  }
```

Chapter 4. Particle Tracking in Discrete Domains

As we have seen, implementation of Lagrangian particle tracking within an analytical flow field is a simple task. The number of known 2D solutions is limited and 3D solutions are few indeed. Besides, we already know where the particles will go from the streamlines. Most practical applications will be discrete domains (i.e., finite element, finite difference, or boundary element) and the flow fields will be the result of a numerical solution, rather than an analytical one. What this means to particle tracking is that the velocity will be known (usually constant) within each element discretely, not continuously over the domain. Furthermore, the velocity will most likely change from one element to the next, as a particle passes through the domain. Keeping track of which element a particle is in will be a much greater task than simply calculating the spatial coordinates.

More importantly, the boundaries between the current element and all adjacent ones, which are where the particle might eventually enter, determine the maximum step length. We don't want to pass on through one or more elements. We want a particle to reach the end of one element and then step over into the next element. Transitioning from one element to an adjacent one has zero spatial length and occurs in zero time. We simply change the element index.

To find the maximum time step, we calculate the intersection with each of the boundaries, taking the one with the smallest positive value. Since the velocity is constant over a single element, there's no point performing some complicated integration (i.e., Runge-Kutta). There's also no point taking a step any shorter than this. The only exception, which we will cover in Chapter 6, is diffusion and dispersion that might be seen as random variations of the velocity within a single element.

The reason most Lagrangian particle tracking codes don't follow this seemingly obvious approach (time steps determined by element size and velocity) is that *snapshots* of the particle field (i.e., position and perhaps concentration) are most easily generated at regular intervals. This is why most Lagrangian particle tracking codes advance all of the particles at the same time (i.e., marching forward simultaneously). In this methodology, every so often you write out a graphic and/or concentration field.

Another reason for programming this methodology is that the memory requirements don't continue to grow as the simulation advances because you only need to store the current values. You can also restart this process where it ended (if you save the files), which fits well with shared and scheduled computer resources and also lends itself to distributed processing. These reasons became superfluous with the ubiquitous availability of microprocessors with gigabytes of RAM—basically with the introduction of the Intel® Pentium® in

1995, about the time I ditched Lagrangian particle tracking in favor of Hamiltonian, which we will discuss in Chapter 5.

For the purposes of illustration (it's not worth putting a lot of effort into what we already know is an inefficient algorithm, especially when an efficient one has been available for 20+ years), we will use the Lagrangian method to track particles through a uniform rectangular grid with a contrived velocity field. This can be readily implemented in 2D and 3D, as the element-to-element transitions are easily calculated. The codes (Lagrangian2.c and Lagrangian3.c) can be found in the online archive in folder examples\Lagrangian. For variety, we can slightly modify the flow field:

```
vars flow(double x,double y)
  {
  static vars t;
  t=FlowPastCylinderWithCirculation(D/6.,-3.*D,x,y);
  t.v+=x-y;
  if(t.r<1./3.)
    t.u=t.v=0.;
  return(t);
  }
```

We first calculate the contrived velocities and store them in arrays, as we might do when loading results from some more elaborate flow model:

```
for(k=i=0;i<ny;i++)
  {
  Y=Ym+(i+1)*(Yx-Ym)/(ny+1);
  for(j=0;j<nx;j++,k++)
    {
    X=Xm+(j+1)*(Xx-Xm)/(nx+1);
    p=flow(X,Y);
    U[k]=p.u;
    V[k]=p.v;
    fprintf(fp,"%lG %lG %lG %lG\n",X,Y,p.u,p.v);
    }
  }
```

Because the grid is uniform, the element-to-element bookkeeping is quite simple:

```
for(s=0;s<max_steps;s++)
  {
  j=(int)((Y[0]-Xm)*nx/(Xx-Xm));
  i=(int)((Y[1]-Ym)*ny/(Yx-Ym));
  if(i<0||i>=ny||j<0||j>=nx)
    break;
  k=nx*i+j;
  RungKutta4(step,&t,time_step,Y,dY,2);
  if(Y[0]<Xm||Y[0]>Xx||Y[1]<Ym||Y[1]>Yx)
    break;
  fprintf(fp,"%lG %lG %lG\n",Y[0],Y[1],t);
```

```
}
```

The 4th-order Runge-Kutta integration again provides smooth particle tracks:

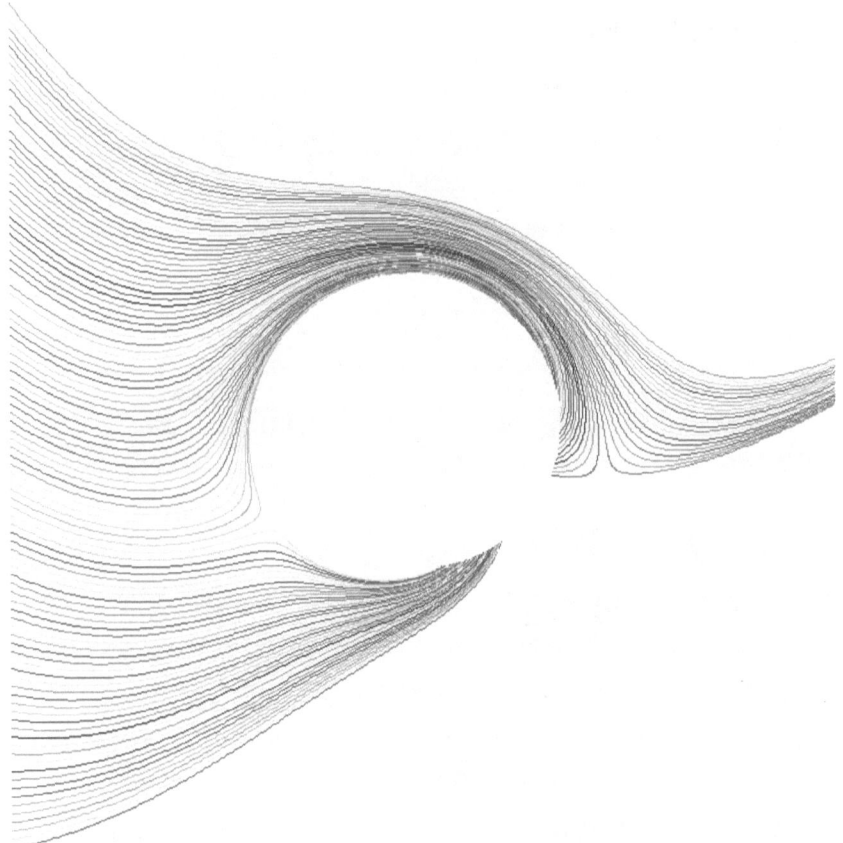

The transition to 3D (Lagrangian3.c) is also quite simple and we can contrive a different velocity field to investigate. For illustration, we create an upward swirling flow field (i.e., a tornado):

```
vars flow(double x,double y,double z)
  {
  double a,r;
  static vars t;
  r=hypot(x,y);
  if(r>FLT_EPSILON)
    a=atan2(y,x);
  else
    a=0.;
  t.u=-sin(a);
```

```
t.v=cos(a);
t.w=1.;
return(t);
}
```

A side view of the particle tracks is:

Viewed from the top down (or bottom up), the tracks are circular:

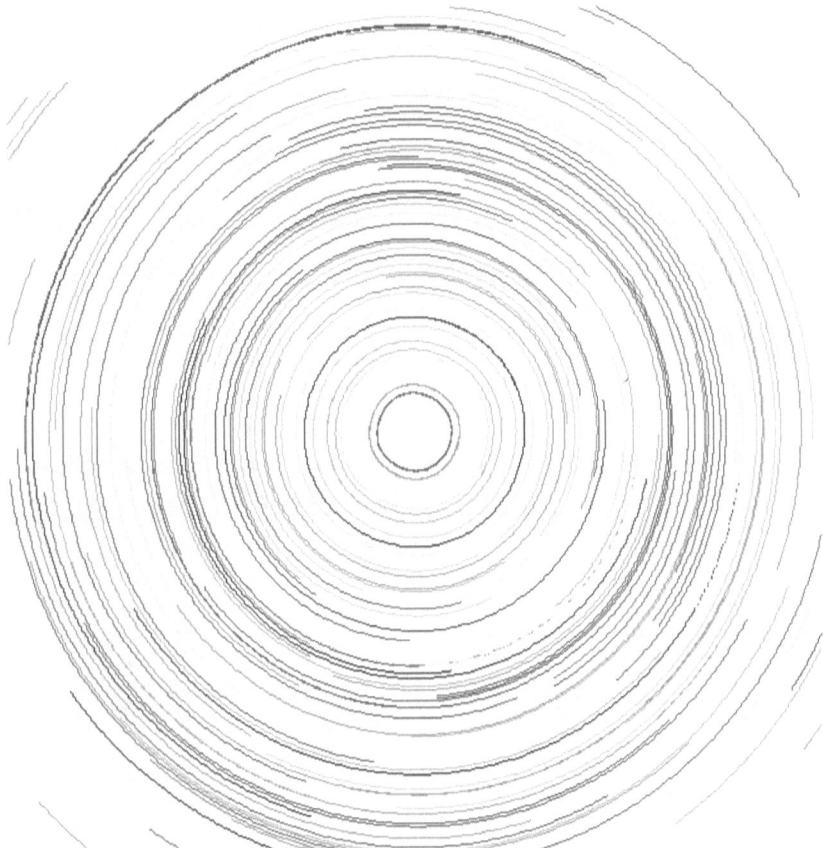

That's it for Lagrangian Particle tracking of simplistic fields and domains. We now move on to realistic problems and practical implementations.

Chapter 5. Hamiltonian Particle Tracking

As briefly mentioned previously, what we are calling the Hamiltonian approach considers position and velocity the independent variables and time the dependant variable. Consider an incremental step in three dimensions:

$$dS = \sqrt{dX^2 + dY^2 + dZ^2} \qquad (5.1)$$

at some velocity magnitude:

$$|\vec{V}| = \sqrt{U^2 + V^2 + W^2} \qquad (5.2)$$

all within a single element. Except for diffusion and dispersion, which we will discuss in Chapter 6, the velocity is constant over the element, so there is no reason not to step completely across it to the adjacent element. The time within this element is then:

$$\Delta t = \frac{\sqrt{dX^2 + dY^2 + dZ^2}}{\sqrt{U^2 + V^2 + W^2}} \qquad (5.3)$$

Implementation

To facilitate calculating intersections, all quadrangles are split into two triangles, prisms are split into three tetrahedra, and bricks (hexahedra) are split into four tetrahedra. Thus, we only need calculate the intersection of a 2D vector with three line segments for each triangle or a 3D vector with three bounded triangular planes for each tetrahedron. The splitting is automatically performed inside the particle tracker to minimize preprocessing.

If snapshots or concentrations are required, these are initialized after reading loading the nodes and elements, then updated for each particle as it passes through each element. If decay occurs, such as with a radioactive component, this is calculated as needed for each particle as it passes through each element. We can even color the tracks based on decay so that they begin with red and proceed through the rainbow to blue.

Flow model results may or may not include porosity, as in groundwater flow through porous media. Porosities and hydraulic conductivities can be read in for each element or assigned by default. Units (SI, English, whatever) don't matter, as long as they are consistent (velocity=length/time). Default porosity (in the .CFG file) can be used to scale the velocities. I began work on the code, PTRAX, in May of 1995 and completed it in March of 1997 with minor features added and a few bugs fixed since then. It has been thoroughly validated against analytical solutions and field tests (see Appendix D) and has withstood the test of time (and court!). While PTRAX began with groundwater applications, it has been successfully used in surface water, atmospheric, branched pipe networks, solids, crystals, and even astronomical simulations. Contaminant animations (which are automatically generated when activated) were first a diagnostic tool, but turned out to be the most compelling feature, especially helpful in court cases.

<u>Two-Dimensional Examples</u>

The fastest way to create complex 2D flow fields is the boundary element method (see Appendix B). The code (PFLOW.c) is in the online archive in folder examples\bem. All you need to get started is a polygon. You can create and edit polygons with AutoCAD®, but a far more convenient tool is my polygon editor, PolyEdit, which can be free downloaded from the web site listed in the Forward. I have modified the BEM code to spit out all the files needed to run the particle tracker. You create the boundary file (see examples *.BEM), run PFLOW, and then run PTRAX. All of the files for a particular project must have the same name but different extensions.[4] The PFLOW input file is: name.BEM.

Table 5.1. File Extensions

recognized file extensions	
extension	contents
name.2DV	generic 2D elements
name.3DV	generic 3D elements
name.BEM	PFLOW input file
name.ELM	elements
name.FPR	element properties
name.NOD	nodes
name.P2D	2D tracks or polygons
name.P3D	3D tracks or polygons
name.TB2	2D tabular fields
name.TB3	3D tabular fields
name.V2D	2D velocities
name.V3D	3D velocities
name.VEF	fracture velocities
name.VEP	porous media velocities

All of these are recognized by my graphics program, TP2[5], which can be freely downloaded from the site listed in the Forward.

---

[4] If you're running Windows®, you really should go to Control Panel, Folder Options, and turn off [x] hide file extensions for known file types, as this is one of the stupidest things Bill Gates ever came up with. Of course you need to see the file extensions and know what they mean, not just the icon of the application that they might be associated with.

[5] I wrote TPLOT in 1980. A coworker took a copy of the source, which was eventually turned into a commercial product without his or my involvement. TP2 is the second generation TPLOT and freely available. It recognizes 27 different file types and runs on any version of Windows®.

The files (beginning with the polygons) can be found in the online archive in the folder examples\bem having names that begin with Gibraltar. The dimensions are nautical miles (i.e., minutes of latitude) with X=0 at Greenwich and Y=0 at the equator. The velocities are arbitrary and can be scaled as desired, since the flow is inviscid. Viscous flow with large-scale turbulence and eddies is beyond the scope of this book. Here we are concerned with particle tracking.

Boundary conditions for the BEM are defined along the polygons (there must be an external polygon and there may be none or several internal ones representing islands). See header at the top of PFLOW.c for details.

Table 5.2. Boundary Condition Types

type	boundary condition
0	U defined
1	dU/dN defined
2	dU/dT defined
3	U defined and end of this boundary
4	dU/dN defined and end of this boundary
5	dU/dT defined and end of this boundary

The top of Gibraltar.BEM is:

```
Strait of Gibraltar
*dimensions are nautical miles (one minute of latitude)
93 -1000
209.0 2206.0 0 1
189.2 2205.0 0 1
169.5 2202.4 0 1
150.9 2196.0 0 1
131.1 2196.0 0 1
111.2 2193.7 0 1
91.4 2192.0 0 1
71.5 2190.0 0 1
52.8 2183.6 0 1
34.1 2176.7 0 1
16.1 2168.1 0 1
2.2 2153.9 0 1
-15.8 2149.9 0 1
-32.5 2146.8 0 1
```

The first line is the title. Any line beginning with an asterisk (*) is a comment. The second line gives the number of boundary points and optionally the number of internal velocities to calculate. Enter a positive number if you intend to supply the points after the boundary or a negative number if you want them automatically distributed uniformly within the interior. The following lines are the boundary (x,y) the type of boundary condition and the value. The first entry is x=209.0, y=2206.0, type=0, value=1.

19

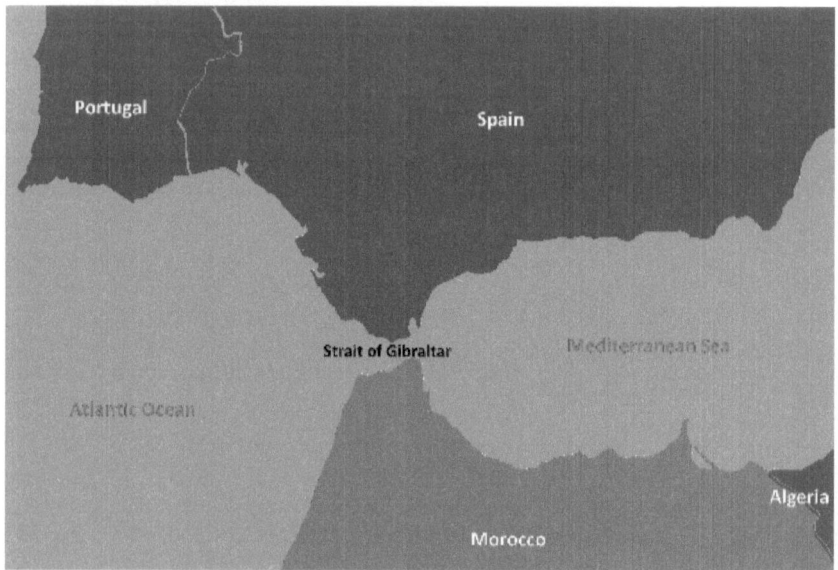

The boundary element results from PFLOW.c are shown below. Again, PFLOW solves for the flow field and generates all of the output files in a few milliseconds. This figure shows velocity vectors on top of the potential field.

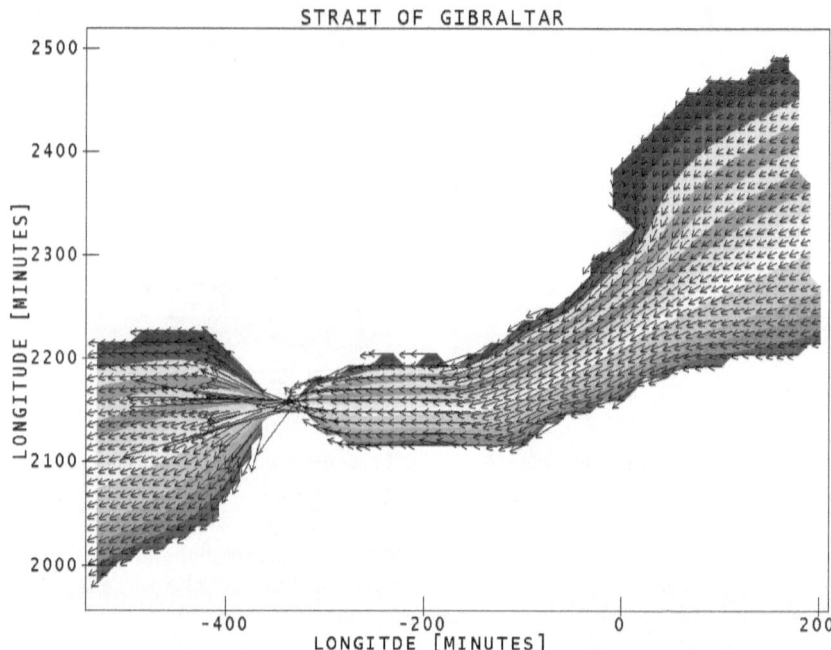

The automatically generated grid of triangles is shown below:

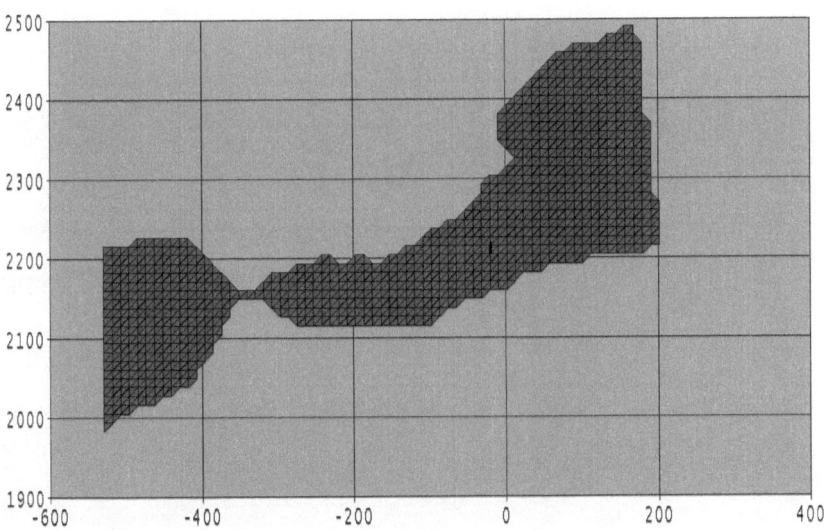

All of the input files are then ready for **PTRAX**, so you just run it. The particle tracks will be displayed and a plot file will also be created.

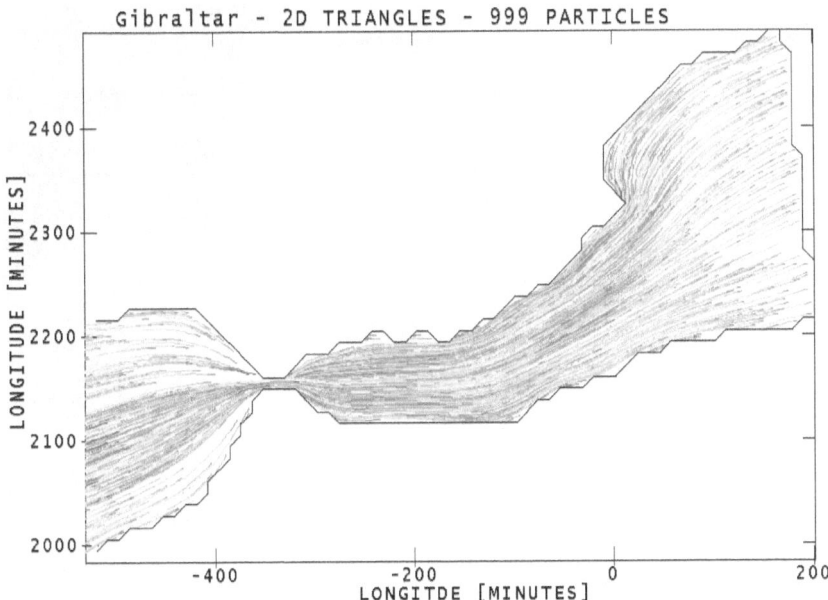

The default is 999 particles randomly seeded throughout the domain. As shown in the log file (PTRAX.LOG), the time required to read the model, generate and track the particles, and create the output files is 3 seconds (on a

21

3GHz Intel® processor). Also by default, the particles age so that they start red and decay through the rainbow to blue.

I already had polygons for the world, including continents, islands, lakes, and countries. I simply cut out this portion using PolyEdit and pasted it onto Gibraltar.BEM, then added 0 0 or 0 1 to define where the potential changes along the north and south boundaries. This Gibraltar example took about five minutes to create and solve. It took longer than that to find a suitable map with Google®.

Strait of Hormuz

The geometry is similar, though the water bodies are quite different. The coordinates are in nautical miles and all of the files are named Hormuz.*.

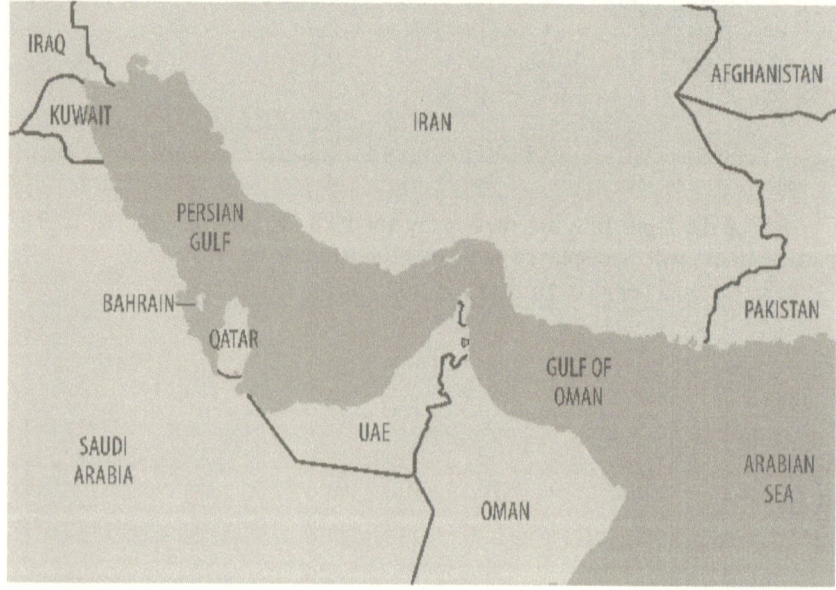

The boundary element input file is also similar:

```
Strait of Hormuz
*dimensions are nautical miles (one minute of latitude)
51 -1000
3030.8 707.7 0 0
2978.1 687.6 0 0
2922.4 678.8 0 0
2866.9 668.0 0 0
2814.3 654.4 0 0
2760.5 646.8 0 0
2711.6 633.6 0 0
2655.7 628.0 0 0
2613.8 669.6 0 0
2580.1 696.0 0 0
```

The velocity vectors and potential are shown in this next figure:

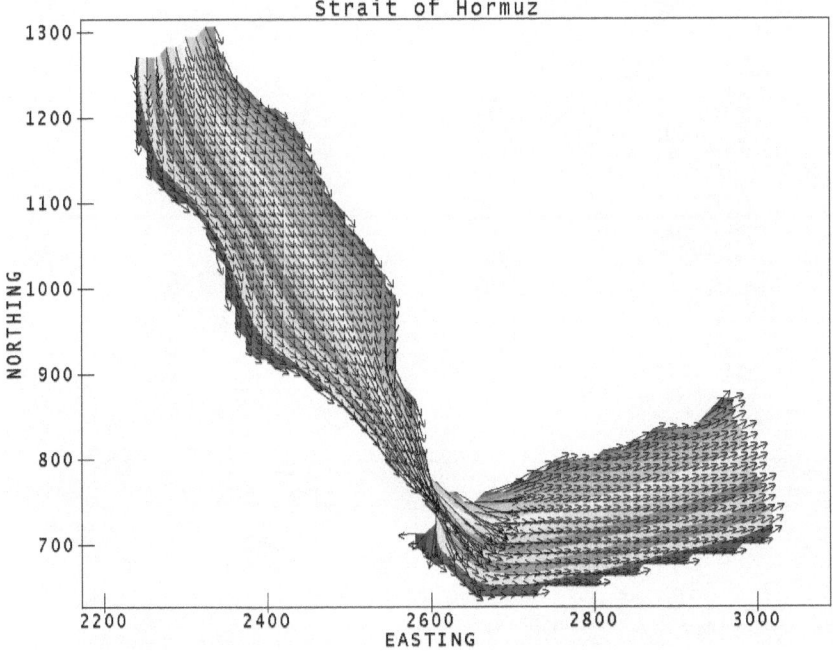

Strait of Hormuz

The default particle tracking parameters are in Hormuz.cfg:

```
999        <- default number of particles to track
36500      <- track duration (days)
365        <- time step (days)
999        <- maximum steps along a single particle track
0.3 1 1    <- default porosity and retardation factor in
     matrix and fractures
0 0 0      <- default longitudinal, lateral, and
     transverse dispersivity in feet
1E30 1E30  <- default matrix/fracture decay half-lives
0.05 0 0   <- default velocities
0 0 0      <- matrix diffusion coefficients
0          <- user-defined synchronous time step (set to
     zero for automatic)
4          <- maximum times a particle can enter an
     element
4          <- maximum times a particle can enter a
     fracture
0 0        <- matrix & fracture tortuosity (0=none,
     1=complete)
```

The PTRAX informational display is shown below:

```
999 particles tracked in 3 seconds                                        _ □ ×
PTRAX/V5.01: Particle Tracker by Dudley J. Benton

animations ....................................... enabled
fractures ........................................ enabled
input format ..................................... Frac3D & ModFlow

application prefix ............................... Hormuz
reading default parameters ....................... Hormuz.CFG
default seeds .................................... 999 particles
track duration ................................... 36500 day(s)
time step ........................................ 365 day(s)
maximum steps along particle track ............... 999 steps
default porosity & retardation factor ............ 0.3,1,1
default dispersivities ........................... 0,0,0 ft
default matrix half-life ......................... 1E30 day(s)
default fracture half-life ....................... 1E30 day(s)
default velocities ............................... 0.05,0,0 ft/day
matrix diffusion coefficients .................... 0,0,0 ft²/day²
synchronous time step ............................ default
maximum times a particle can enter an element .... 4
maximum times a particle can enter a fracture .... 4
matrix,fracture tortuosities ..................... 0,0
concentrations divided by porosity ............... NO
stray particles will be ignored .................. OK
trapped particles will be ignored ................ OK
include empty elements in snapshot files ......... OK
create track file ................................ OK

input model ...................................... FRAC3D
FRAC3D node file: Hormuz.NDE ...................... 996 nodes
range of X ....................................... 2241.13≤X≤3017.47
range of Y ....................................... 640.13≤Y≤1307.3
FRAC3D element file: Hormuz.ELM ................... 1804 elements
elements composed of 3 nodes ..................... TRIANGLES
reading element nodes ............................ 1804 elements
node:element links ............................... 5412 links
element:element links ............................ 5226 links
model boundaries ................................. 186 faces
computing element areas .......................... 1804 elements
grouping elements ................................ 437 groups
FRAC3D velocity file: Hormuz.VEP .................. 1804 velocities
properties ....................................... default
transport properties ............................. default
dispersion ....................................... OFF
diffusion ........................................ OFF
scattered seeds .................................. 999 seeds
particle track file .............................. Hormuz.TRK
time to track particles .......................... 1 seconds
net performance .................................. 59940 seeds/minute
tracks ended at boundaries ....................... 945 seeds
tracks ended due to maximum time ................. 54 seeds
track plot command file .......................... Hormuz.TP2
total elapsed time ............................... 3 seconds
3 notes + 0 warnings
for summary see log file, PTRAX.LOG
```

It took all of 3 seconds to read in the files and track 999 particles.

The **PTRAX** on-screen particle tracks are shown in this next figure:

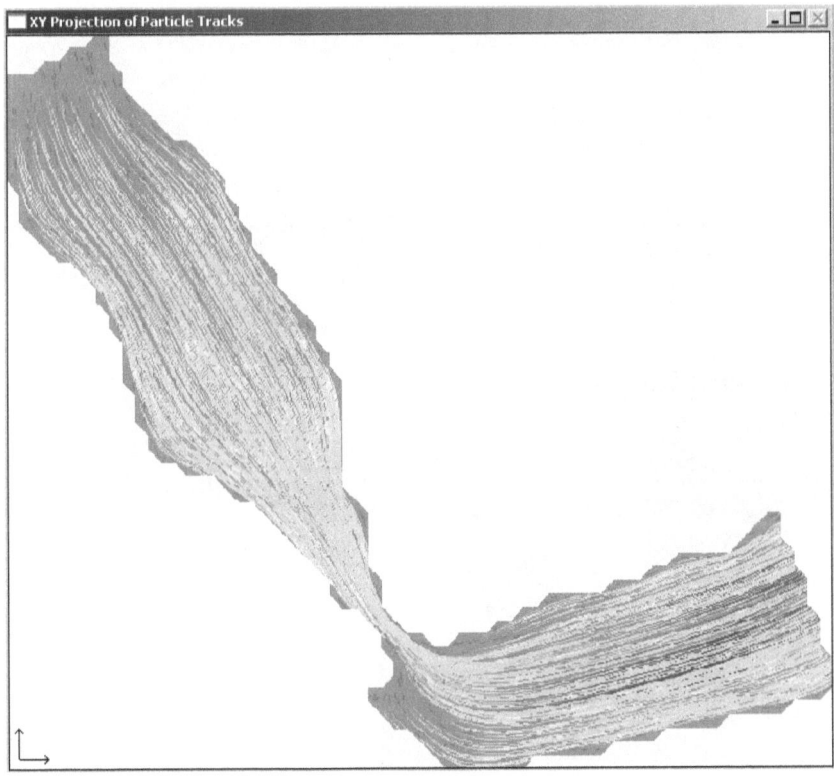

XY Projection of Particle Tracks

Dardanelles

The Dardanelles, also known from Classical Antiquity as the Hellespont, is a narrow, natural strait and internationally significant waterway in northwestern Turkey that forms part of the continental boundary between Europe and Asia, separating Asian Turkey from European Turkey. One of the world's narrowest straits used for international navigation, the Dardanelles connects the Sea of Marmara with the Aegean and Mediterranean Seas, while also allowing passage to the Black Sea by extension via the Bosphorus. The boundary element input file begins with the following:

```
Dardanelles
*the strait that divides Europe from Asia Minor (i.e.,
   Turkey)
*dimensions are nautical miles (one minute of latitude)
56 -1000
1770.7 2441.1 0 2
1760.8 2440.6 0 2
1751.0 2439.5 0 2
1741.3 2437.9 0 2
```

25

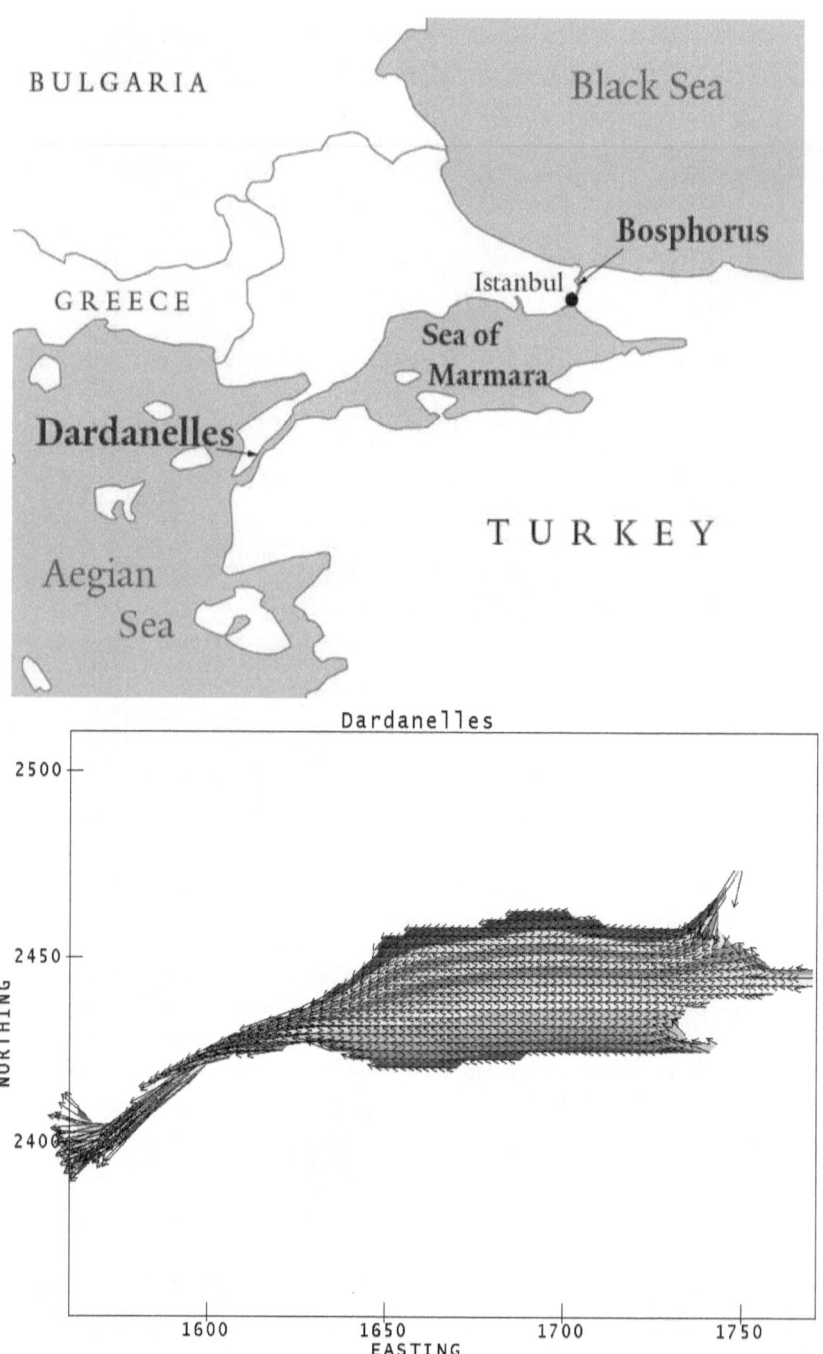

The default particle tracks are:

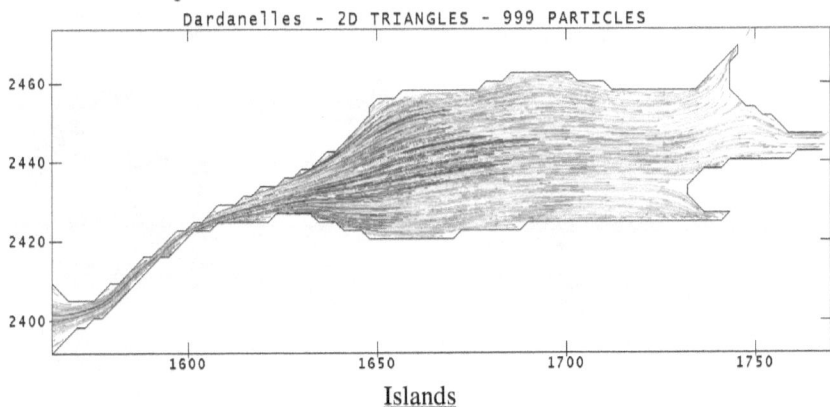

Dardanelles - 2D TRIANGLES - 999 PARTICLES

<u>Islands</u>

The boundary element method can also handle islands. Note the "end of this boundary" designations in Table 5.2. There is no limit to the number of polygons used to describe the domain, although the BEM does eventually require solution of simultaneous linear equations and this is definitely limited. The number of equations is roughly twice the number of boundary points.

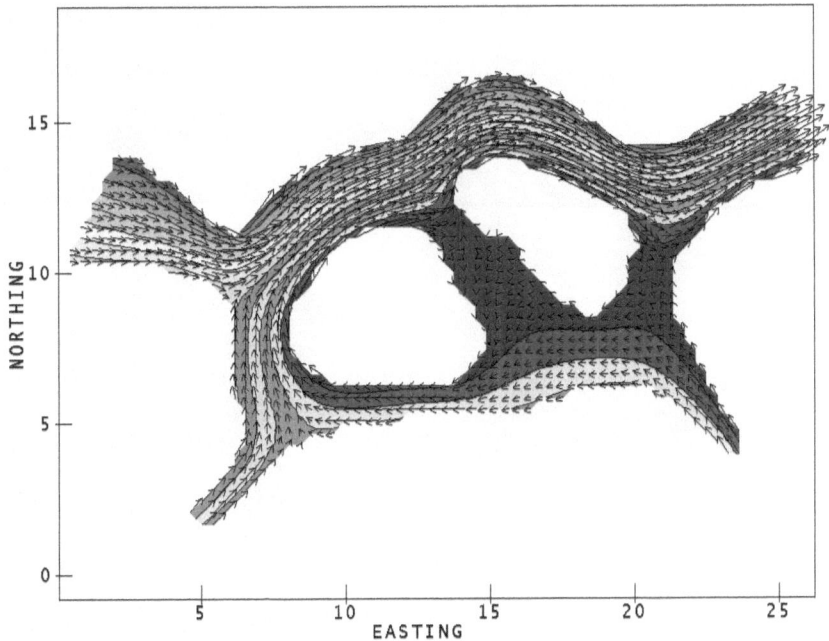

The default particle tracks are:

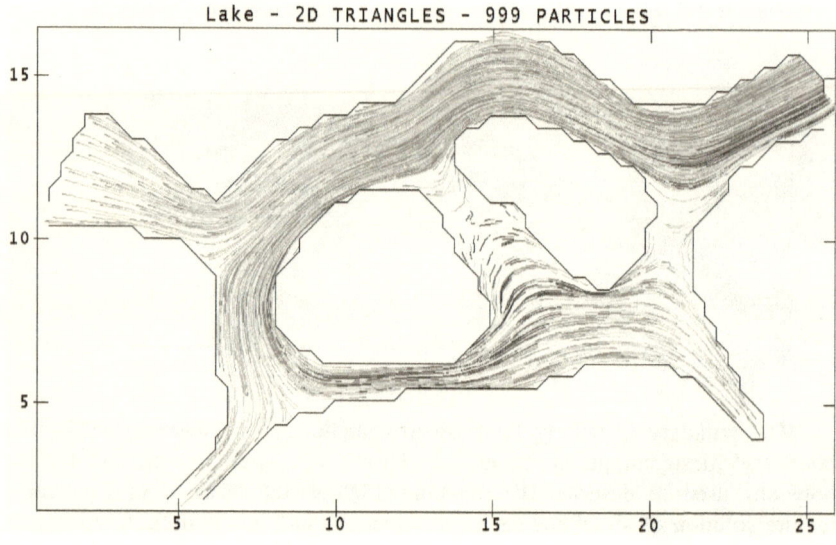

Lake - 2D TRIANGLES - 999 PARTICLES

Chapter 6. Diffusion and Dispersion

In the context of particle tracking, diffusion is random movement of the particles through the domain over time, irrespective of the velocity field or even in the absence of any velocity. This is not just any random movement. There are specific quantitative expectations. Diffusion is well-defined analytically and diffusion coefficients, derived from experiment, have been published for many years. If enough particles are tracked, the result must match analytical solutions where these are available. How many particles are enough? The validation runs described in Appendix D were done using 800,000 and the agreement is excellent. The following example tracked 50,000 particles, which was ample to illustrate the process.

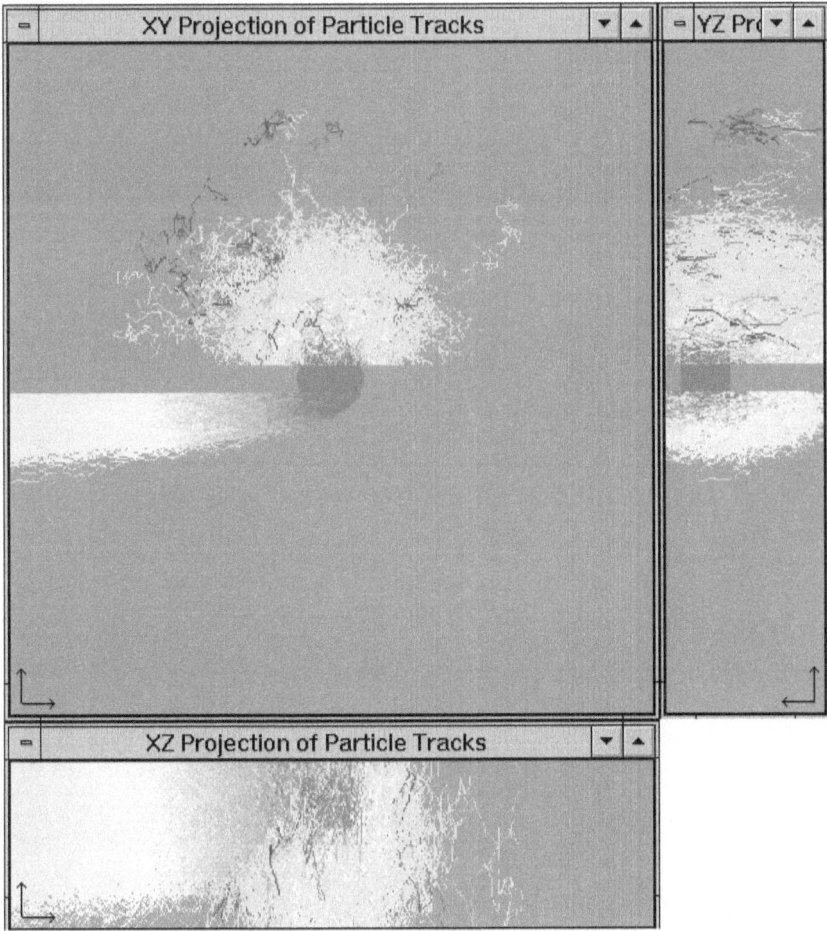

Dispersion is very similar to diffusion, in that it is random movement, but dependant on (i.e., proportional to) local velocity. Experimentally measured dispersion coefficients have also been published and so there are also definite expectations for dispersion. If a particle tracker doesn't meet these expectations, then it's just making up numbers and holds forth no real promise of predictive capability, which is essential for effective remediation and design. The preceding figure illustrates the difference between diffusion and dispersion, both within a single model.

The red circle (actually a vertical cylinder in this 3D model) in the middle of the domain is where the xxx particles were seeded. The white horizontal slot is a section of elements with no dispersion or diffusion. These particles are *stuck* and never move. Elements in the upper portion have zero velocity and non-zero diffusion coefficients. Elements in the lower portion have uniform velocities to the left, non-zero dispersion coefficients, and zero diffusion coefficients. The particles age so that they start out red, proceed through the rainbow, and end blue. The ones in the center along the stagnant slot remain red because they are trapped in the elements and aren't tracked. PTRAX generates four-dimensional results (three spatial dimensions plus time) in the form of concentration fields that can be sliced and also contoured using Tecplot® or TP2. The plan view contours (looking from the top down) are:

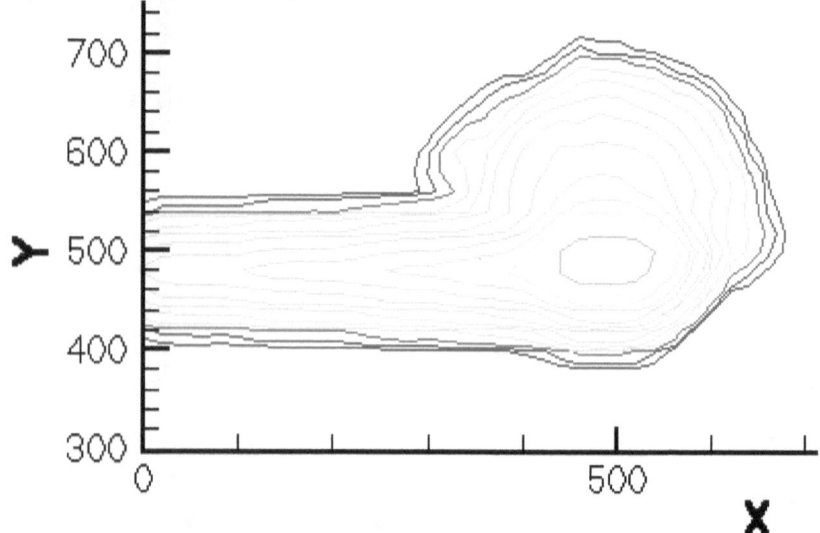

Three-dimensional contours are more accurately described as *shells* of constant value.

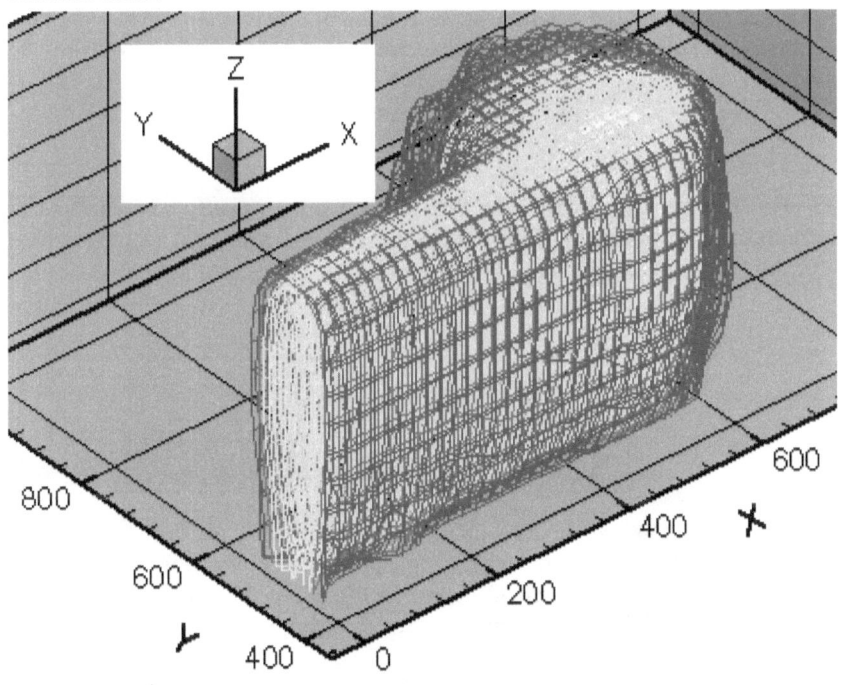

<u>Random Walk</u>

In order for particle tracking results to be meaningful, they must match analytical results. The stepping process used is called a *random walk*, and is often compared to a drunken sailor staggering back to the ship after a long night at the pub. While the final equation seems so very simple, the process of arriving at it in 1995 was anything but simple. Many formulations were tried, tedious simulations performed, results compared, and approaches discarded before arrive at one that was adequately consistent. The random step length associated with a dispersion length is defined by the following equation:

$$\Delta S = R \sqrt{2\,\alpha\,|\vec{V}_m|\,\Delta T} \qquad (6.1)$$

where ΔS is the random step length, R is a normalized random number (i.e., having a mean of 0 and a standard deviation of 1), α is the dispersion length, $|V_M|$ is the magnitude of the mean (i.e., non-random) velocity, and ΔT is the time step. Diffusion is handled in this same way, except that the dispersion length times the magnitude of the vector velocity is replaced by the diffusion coefficient, D.

For dispersion (or diffusion) in several directions, multiple random numbers (i.e., Rs) and directionally associated dispersion lengths (i.e., α_X, α_Y, α_Z) are combined to form the random steps (i.e., ΔS_X, ΔS_Y, ΔS_Z). For a particle traversing a cell, there is an effective random velocity associated with the random step length and implied time step.

$$\left|\vec{V}_R\right| = \frac{\Delta S}{\Delta T} \tag{6.2}$$

For dispersion in several directions, the effective velocity components can be represented by a mean and random part:

$$U_T = U_M + U_R = U_M + \frac{\Delta S_X}{\Delta T} \tag{6.3}$$

$$V_T = V_M + V_R = V_M + \frac{\Delta S_Y}{\Delta T} \tag{6.4}$$

$$W_T = W_M + W_R = W_M + \frac{\Delta S_Z}{\Delta T} \tag{6.5}$$

where U, V, and W are the velocity components in the X, Y, and Z directions, respectively. If dispersion lengths are specified along the longitudinal, horizontal-transverse, and vertical-transverse directions, the corresponding steps along the principle axes are computed using standard trigonometric relationships.

Statistical Requirements

For a statistically large sample (i.e., many particles), the net influence of the random walk on the ensemble of particles must exhibit several properties:

1) The spreading (over that without dispersion) in the direction associated with each α is proportional to the square-root of α and ΔT.

2) The net displacement of the particles (compared to that without dispersion) is zero.

3) The net movement of the mass-weighted centroid of the particles is the same with or without dispersion.

Given these properties and the relationships between the random step length, mean and random velocity components, and time steps, the following requirements can be deduced:

$$\sum_{i=1}^{n} \Delta S_i \approx 0 \tag{6.6}$$

$$\sum_{i=1}^{n} \frac{\Delta S_i}{\Delta T_i} \approx 0 \tag{6.7}$$

32

These summations must hold for a single particle as well as for the ensemble, and they must hold in each dimension. In order to simultaneously satisfy these pairs of relationships, the time steps must be equal (i.e., if they are equal, then ΔT can be brought outside the summation).

Because these statistical relationships require equal time steps, an immediate problem arises, regardless of whether conventional Lagrangian particle tracking or the Hamiltonian method is used. Efficient implementation of a Lagrangian method requires a dynamically adjusted step length. Implementation of the Hamiltonian scheme results in time steps varying over orders of magnitude, as a particle may pass through a cell near a vertex and cover a small distance in a correspondingly small time.

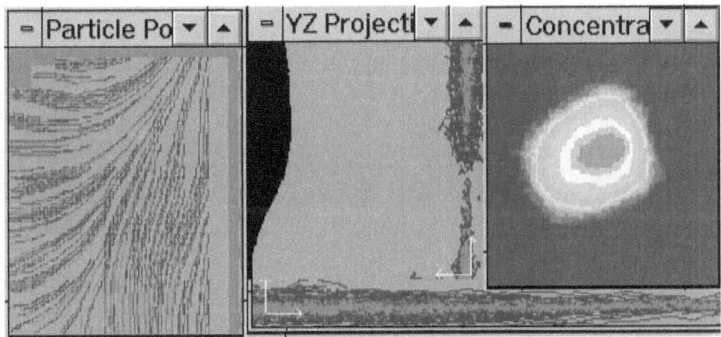

If a constant time step is required, then the smallest required time step becomes a limiting factor and results in impractical runtimes. It is for this reason that the random walk is not often used in large particle tracking applications or such applications are run on super computers. Rather than using such a brute force approach, PTRAX steps around this problem by *joggling* the particles and *quantizing* the time steps. The resulting algorithm is very fast and also accurate, as shown in Appendix D.

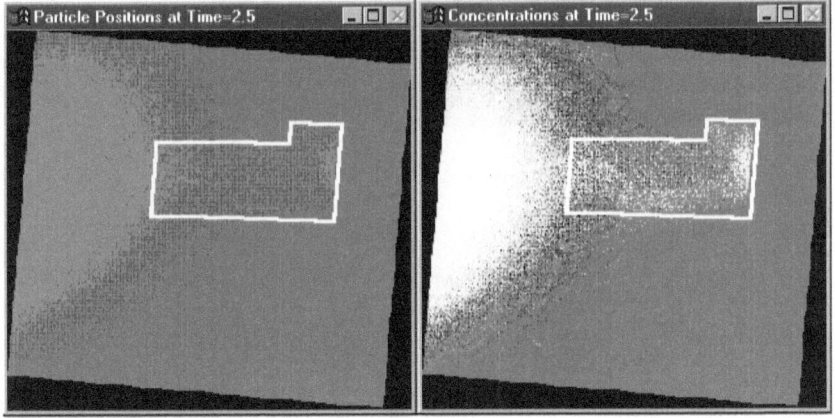

Pure Diffusion

Pure diffusion in 3D results in spreading of the substance, as approximated by the particles, in this case 50,000.

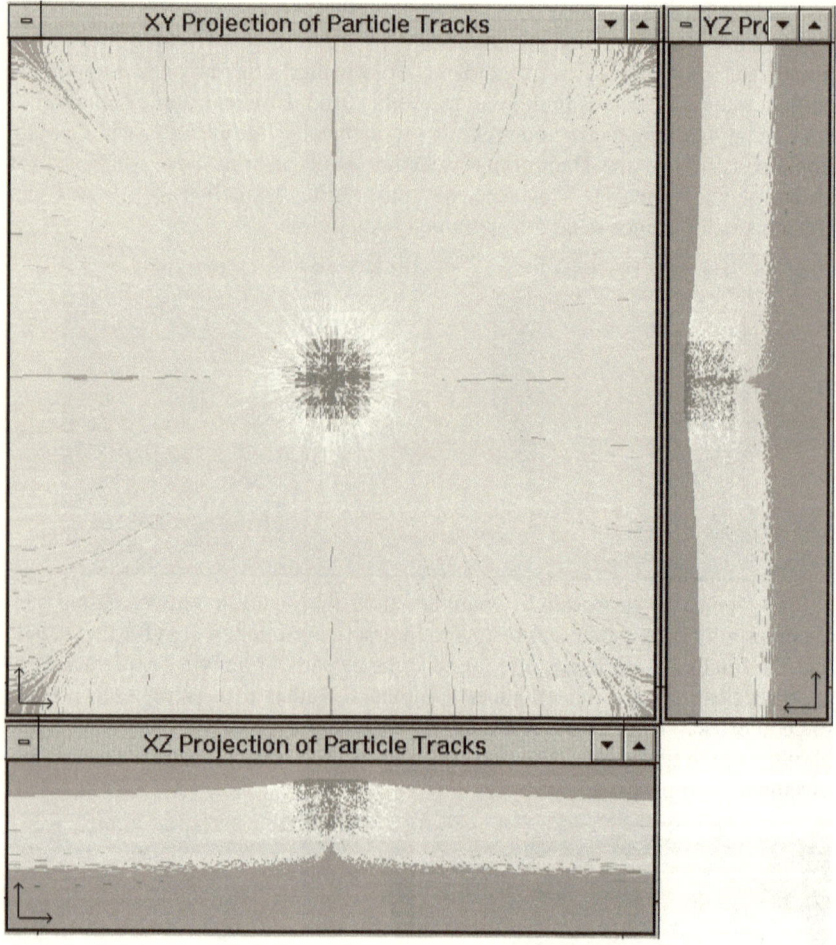

Modified Diffusion

When dispersion is added, even in the presence of uniform velocity, the tracks take on a more random pattern, which shows that PTRAX can handle both types of displacements using a random walk algorithm.

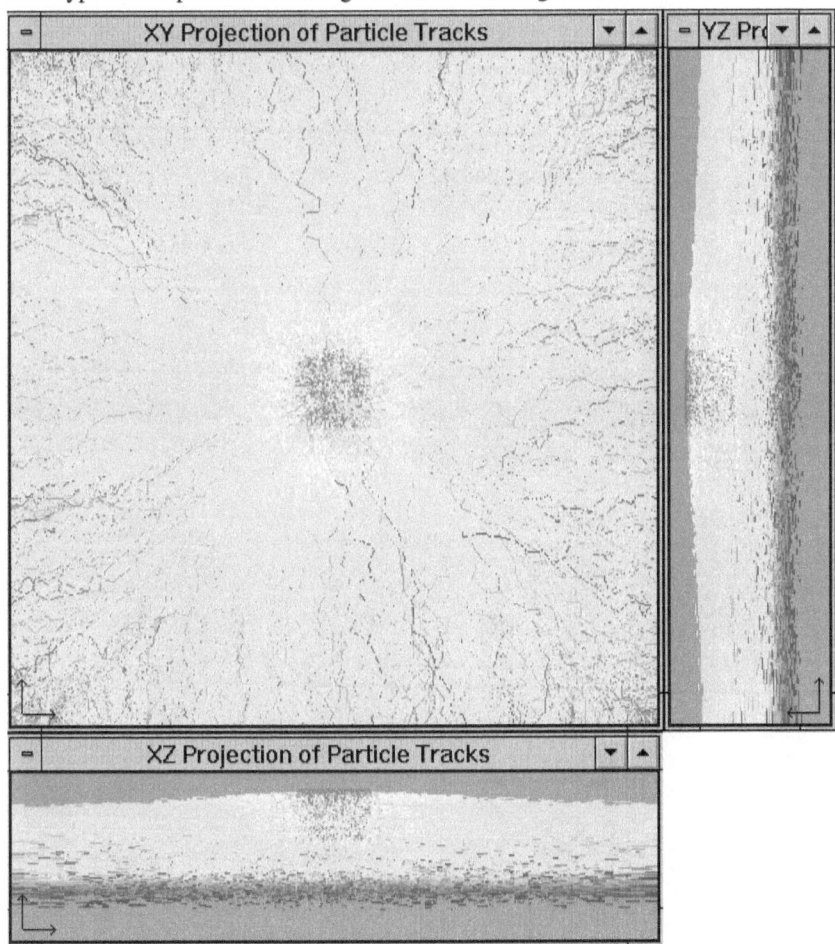

Capture Wells and Walls

In order to accurately evaluate potential remediation strategies, it is necessary to implement wells and walls. Wells are straight-forward enough, but walls are usually deep, narrow trenches lined with special cloth and filled with gravel, like a French drain. Both are commonly installed to inject and/or withdraw water from the ground and process it to remove contaminants. PTRAX handles wells that inject (red) and withdraw (cyan) as well as walls. The robust algorithms handle any sort of boundary, not just smooth domains.

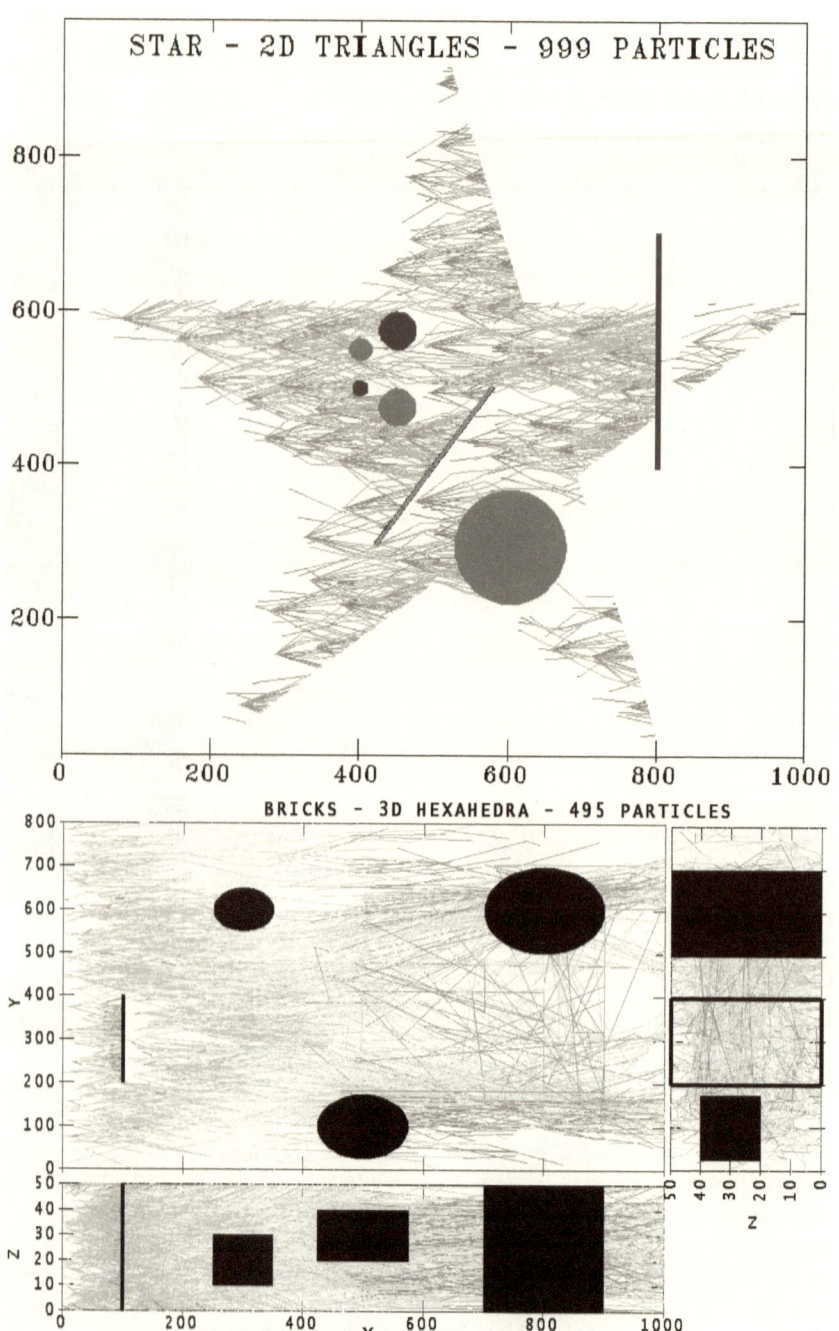

STAR - 2D TRIANGLES - 999 PARTICLES

BRICKS - 3D HEXAHEDRA - 495 PARTICLES

Once the programming logic was refined, any number of complex 3D obstructions can be handled, as illustrated in the previous figure.

2D Dispersion Example

The following example of dispersion in a two-dimensional surface water flow can be found in the online archive in folder examples\bem in files with names beginning with "Reservoir."

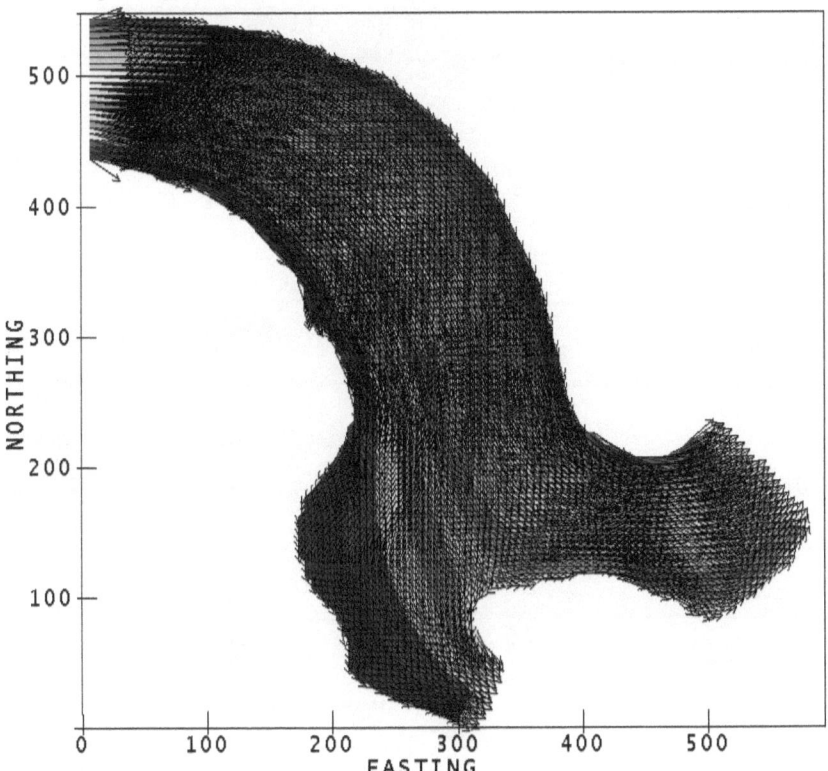

The particle tracks with dispersion are:

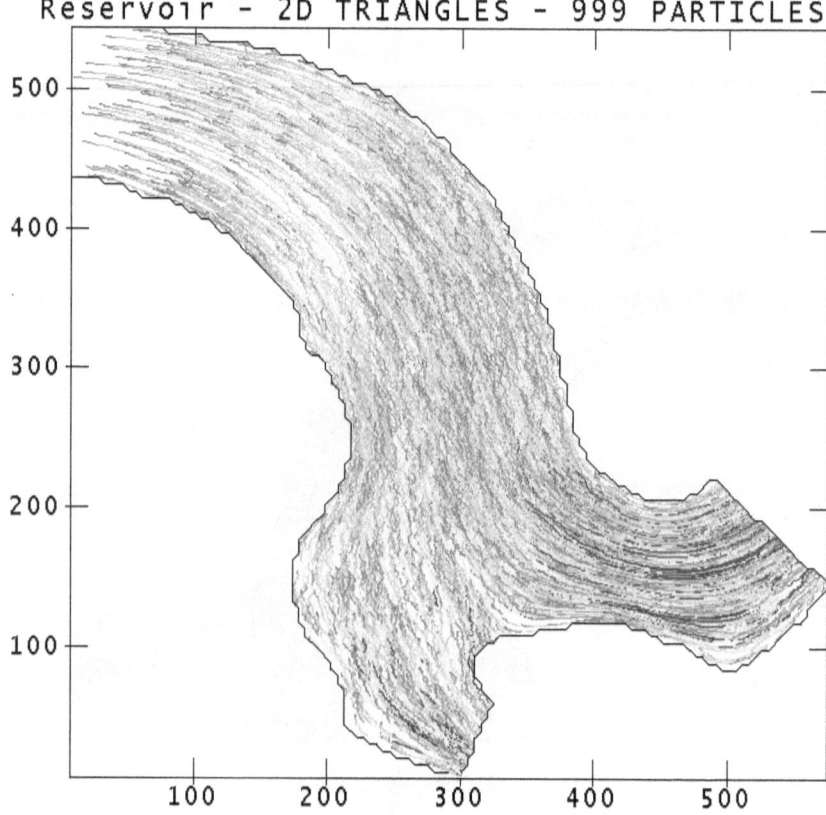

Reservoir - 2D TRIANGLES - 999 PARTICLES

3D Dispersion Example

This first figure shows the particle tracks without dispersion:

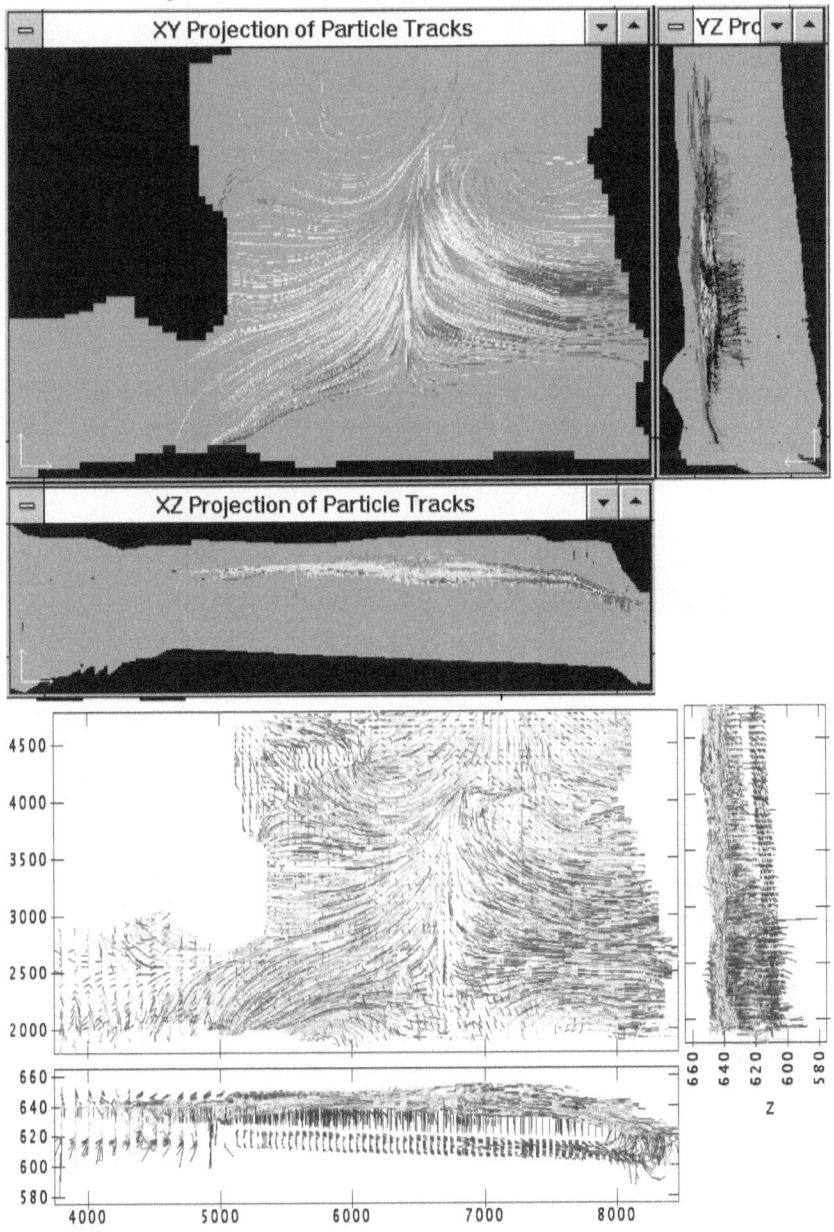

The second figure shows the particle tracks with dispersion:

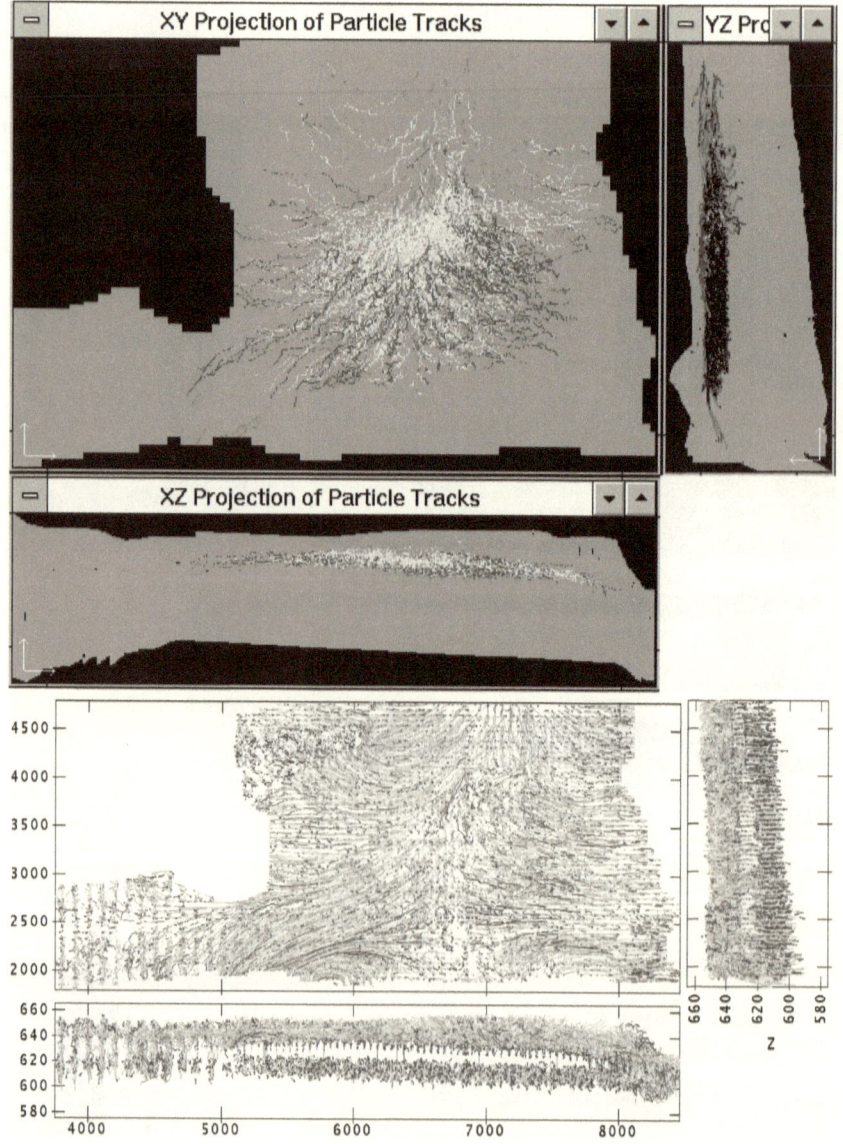

A different 3D domain without diffusion or dispersion:

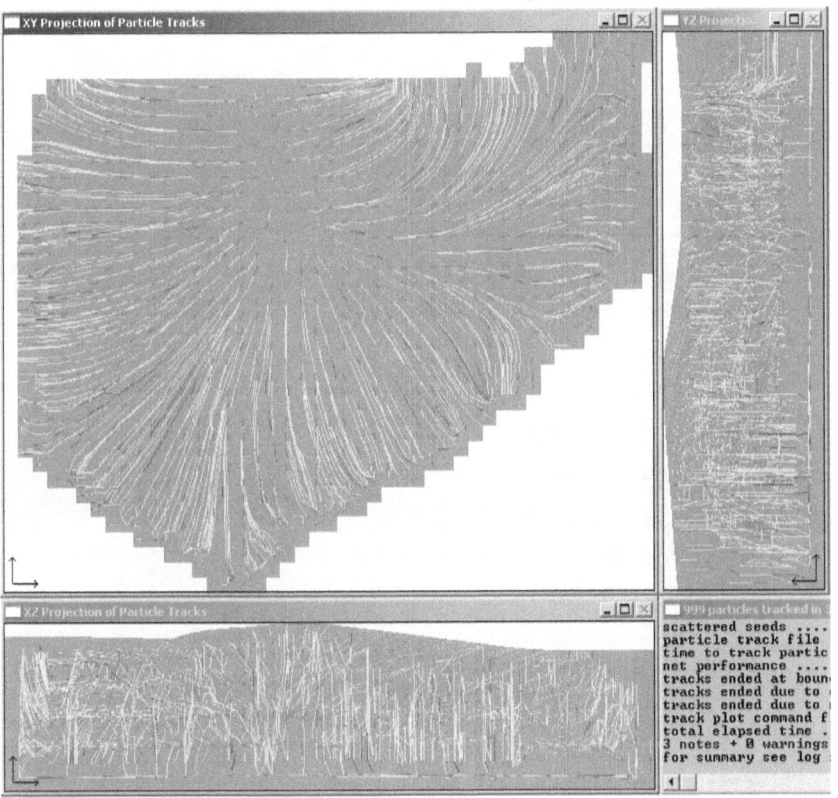

Typical information in the log file:

```
PTRAX/V5.01: Particle Tracker by Dudley J. Benton
animations:    enabled
fractures:     enabled
input format:  Frac3D or ModFlow
parameter 2:   track
parameter 3:   wait
parameter 4:   ok
application prefix: wine
checking for existing output files
  none found
reading default parameters from file: wine.CFG
default parameters
  default seeds: 999
  track duration: 500 days (1.4 years)
  time step: 100 days (0.3 years)
  maximum steps along particle track: 999
  default porosity: 0.3
```

```
      default retardation factors: 0.5.1
      default dispersivities: 3,3,0.3 ft
      default matrix half-life: 1E+030 day(s)
      default fracture half-life: 1E+030 day(s)
      default velocities: 0.05,0,0 ft/day
      matrix diffusion coefficients: 1,1,0.01 ft²/day²
      synchronous time step default
      maximum times a particle can enter an element: 99
      maximum times a particle can enter a fracture: 99
      matrix tortuosity (0=none, 1=complete): 0
      Note: matrix tortuosity OFF
      fracture tortuosity (0=none, 1=complete): 0
      Note: fracture tortuosity OFF
   reading MODFLOW files
   Note: concentrations will NOT be divided by porosity
   stray particles (outside domain) will be ignored
   trapped particles will be ignored
   include empty elements in snapshot files
   create track file
   MODFLOW basic input file: wine.BAS
      Title1: PREFIX: WINE
      Title2: CREATED BY BUILD3D/V1.71
      grid: 45x37x5
      total cells: 8325
      active cells: 5860
      resulting nodes: 10488
      element type: HEXAHEDRA
   MODFLOW block-centered flow file: wine.BCF
      06X67889.85
      06Y66500.16
      145.916Z6235.105
   linking domain
      node:element links
      66600 links
      element:element links
      45800 internal faces
      4150 external faces (boundaries)
      3810 external elements
      4152 external nodes
      computing element volumes
   grouping elements
      largest element: 175.33x175.68x24.485
      27x22x2=1188 groups
      group size: 303.456x309.531x89.195
      sorting groups
      indexing groups
      there are 546 active and 642 empty groups
      the smallest group is 56, containing 5 members
      the largest group is 1, containing 20 members
```

```
the active groups contain an average of 15 members
MODFLOW binary flow file: wine.CBB
characteristic parameters
  length threshold = 1.0223 feet
  volume threshold = 0.00457439 ft^3
  time threshold = 0.0630275 day(s)
  velocity threshold = 1.62199E-005 ft/day
  mean velocity = 16.2199 ft/day
  mean time to traverse element = 5.92921 day(s)
  synchronous time step = (automatic)
random walk
  dispersion ON
  diffusion ON
scattered seeds: 999
  particles tracked: 999
  time to track particles: 2 seconds
  average particles tracked per minute: 25409
  average steps per particle: 126
  average particle track: 887.037
  average particle life: 6911.78
  average particle speed: 0.128337
  total   particle movement: Sp=886150
  random particle movement: Rx/Sp=-0.00255429
  random particle movement: Ry/Sp=0.00155633
  random particle movement: Rz/Sp=-0.00166988
  random time steps: 2.08967 ñ 5.93962 day(s)
  tracks ended at boundaries: 621
  tracks ended due to circulation: 20
  tracks ended due to maximum steps: 63
  tracks ended due to maximum time: 295
Summary of Particle Tracking by Centroid of Mass
```

snap	year	mass	%mass	Xcentroid	Ycentroid	Zcentroid	Xradius	Yradius	Zradius
1	0.0	2.83E-05	100	3560.7	3602.9	183.3	2056.4	1557.5	21.8
2	0.3	1.44E-05	50.8509	3499.2	3196.1	173.6	2299.1	1987.8	18.5
3	0.5	1.21E-05	42.7427	3482.7	2948.1	172.3	2315.0	2077.1	19.0
4	0.8	1.14E-05	40.1401	3516.8	2902.4	170.9	2341.6	2094.8	18.7
5	1.1	1.10E-05	38.8388	3517.2	2890.5	170.5	2344.4	2109.3	18.7
6	1.4	1.07E-05	37.8378	3477.3	2863.4	171.0	2340.7	2109.7	19.4

The configuration file contains pertinent modeling information:

```
999       <- default number of particles to track
500       <- track duration
100       <- time step
999       <- maximum steps along a single particle track
0.3 0.5   <- default porosity and retardation factor
3 3 0.3   <- default longitudinal, lateral, and
   transverse dispersivity in feet
1E+030    <- default decay half-life
```

```
0.05 0 0 <- default velocities
1 1 0.01  <- default matrix diffusion factors
0          <- user-defined synchronous time step (set to
    zero for automatic)
99          <- maximum times a particle can enter an
    element
99          <- maximum times a particle can enter a
    fracture
0          <- scattering factor (0=no scattering,
    1=complete scattering)
```

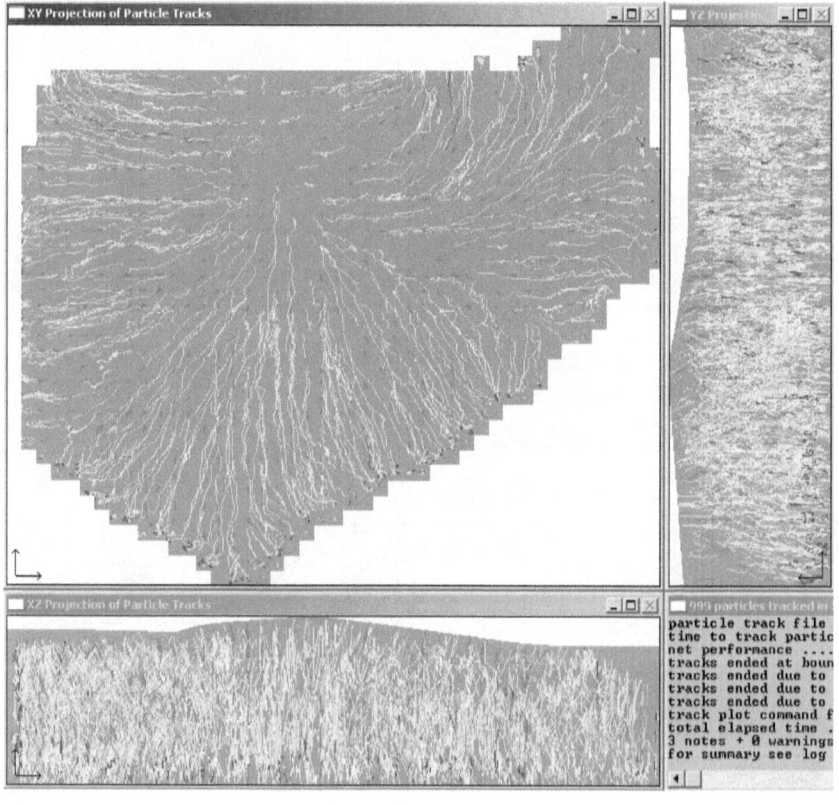

Chapter 7. Flow in Fractures

Fractures are like short-circuits in porous media. Fractured rock is often called *karst*, especially when limestone-based. In modeling, fractures are a 2D network imposed on top of a 3D domain, typically having much higher velocities, as would be the case in actual Earth formations. Particles can enter factures, pass quickly along them, and then return to the porous media at another location, sometimes far away. I devised a series of test cases to demonstrate that the coding works.

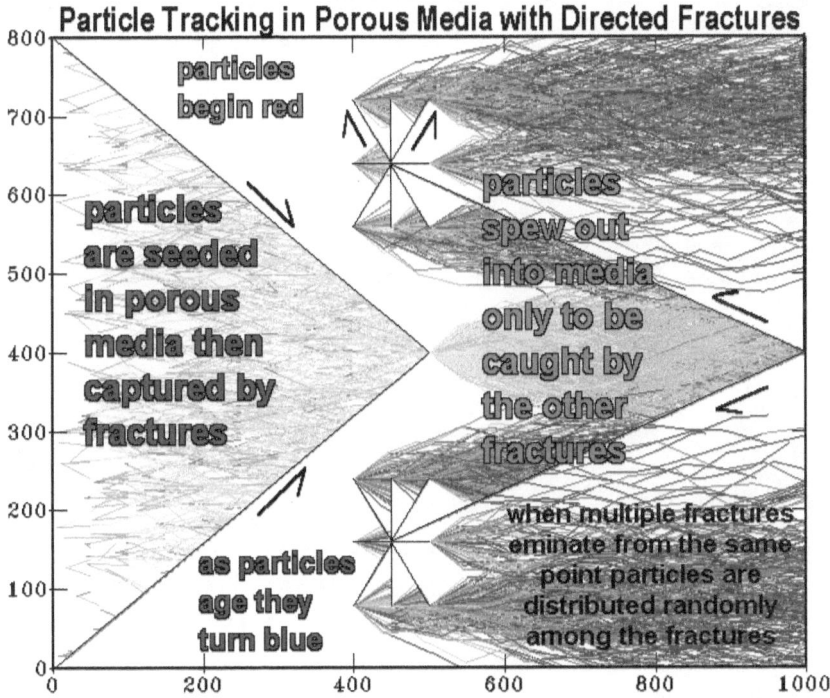

Particle Tracking in Porous Media with Directed Fractures

particles begin red

particles are seeded in porous media then captured by fractures

as particles age they turn blue

particles spew out into media only to be caught by the other fractures

when multiple fractures eminate from the same point particles are distributed randomly among the fractures

In the preceding example, particles are seeded on the left and flow toward the right. When they encounter the first two long diagonal fractures, the particles are trapped and quickly flow to the point, where they spew back out into the porous media. They continue flowing right and begin to spread out, then are caught by the second two long diagonal fractures. These second pair of factures flow back to the left, so the particles run back to the two stars, which consist of eight fractures. When multiple fractures emanate from the same point, the particles are randomly distributed between the possible paths, flow along these, and eventually back into the porous media, where they are again swept to the right. Some escape the right boundary, while others are caught by the fractures again and loop back through again. The particles age along the way from red to blue so that you can tell which ones have been around more than once. The algorithm is robust enough to traverse a 3D maze:

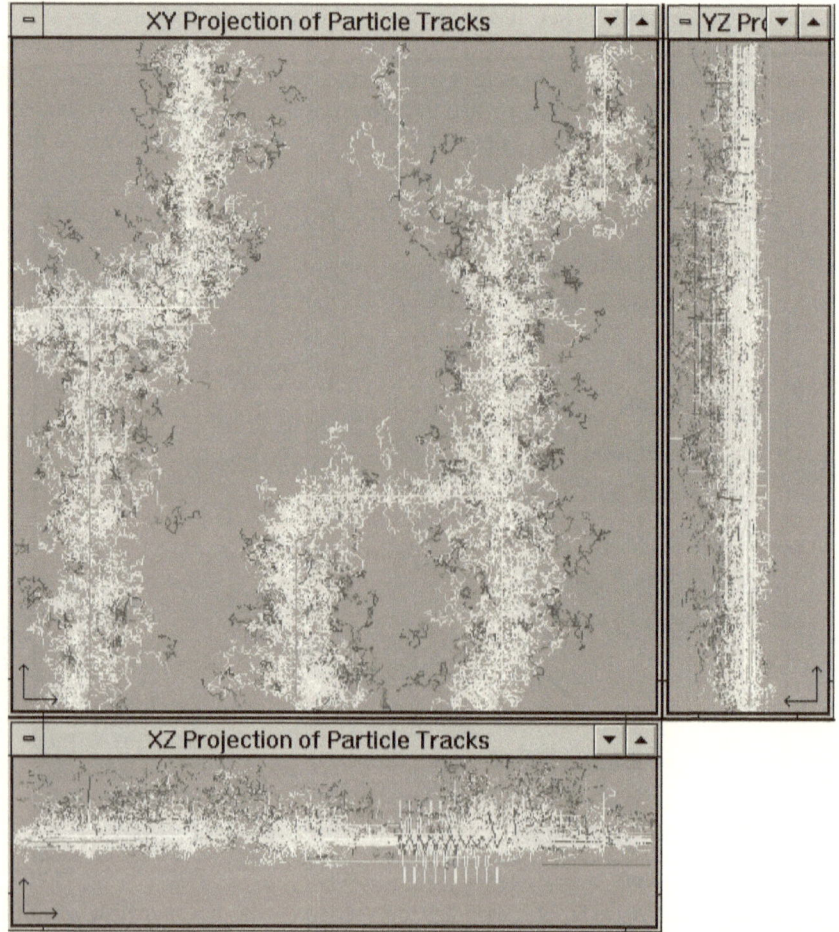

This next figure shows the concentrations:

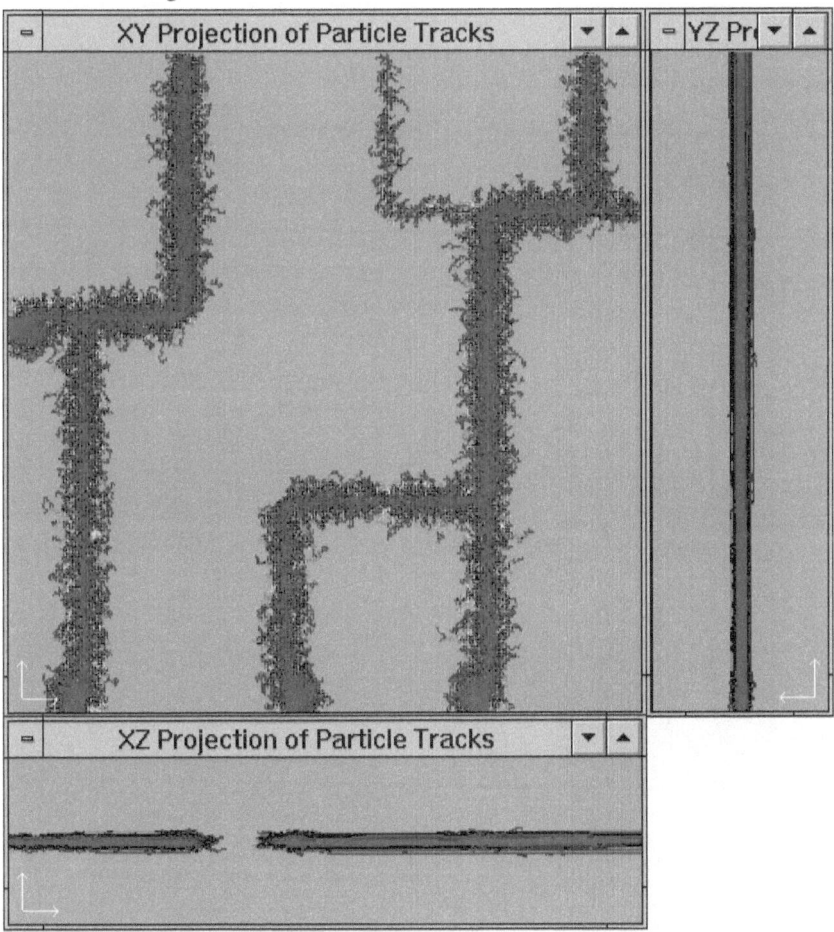

This representation loses a lot, being static. The original (which I still have) is animated, as is the analytical solution on the next page. The files for both of these examples can be found in the online archive in folder examples\fractures.

This next figure is as close as we could get with an analytical solution. Needless to say, setting up the particle tracking was much easier than building the analytical solution, which we accomplished with finite elements and superimposed domains.

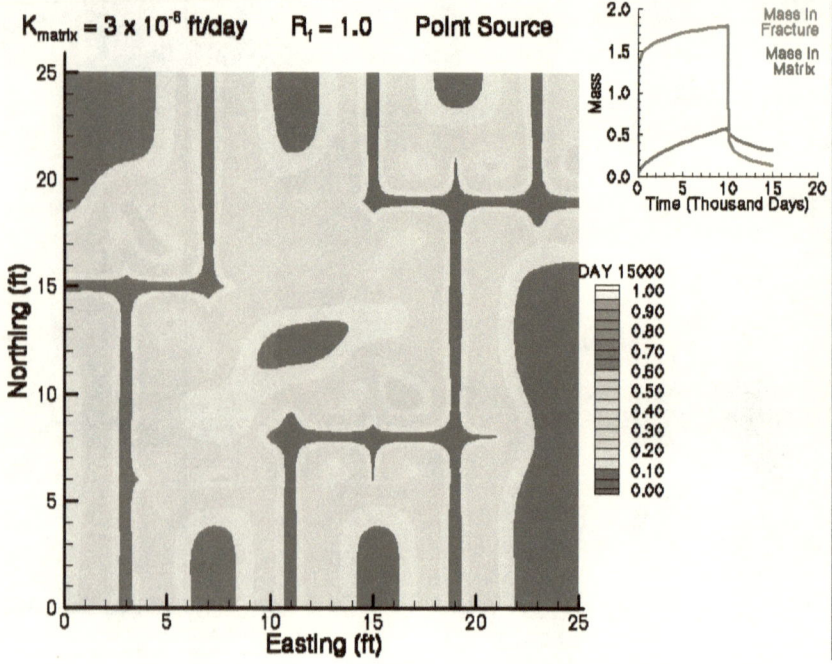

Chapter 8. Contaminant Plumes

Contaminant plumes were the motivation for developing the software in the first place. Somebody put something in the ground that they shouldn't have and now it must be tracked, contained, and removed, which is the purpose of remediation. Contaminant plumes are often the starting point for particle tracking and I have devoted years to constructing these. Whether in the ground or a body of surface water, you will never be able to measure it all. Only point measurements are available plus sometimes a total spill amount. A volume of variable concentrations must be inferred from the point measurements. This example is typical of the information available, except that there were far more data points than are often available. The thick magenta curve represents the estimated boundary. The red curves represent surface features, including roads and a parking lot. The data points are colored based on the log of the concentration of the contaminant.

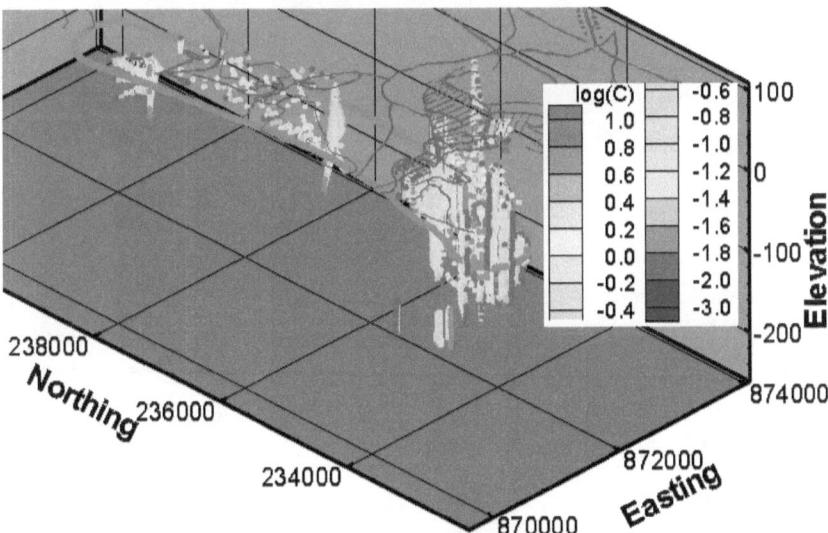

In groundwater, data points are most often a vertical sequence, as these are withdrawn from wells. The contaminant plume must be defined for the entire initially affected zone. This requires specialized software and interpolation strategies, including: linear, inverse distance, and kriging.[6]

---

[6] Used mainly in geostatistics, kriging or Gaussian process regression is a method of interpolation in which the continuously estimated values are approximated by a Gaussian relationship. In some cases, kriging gives the best linear unbiased prediction of the continuous values. Interpolating methods based on other criteria (such as smoothness) may not yield the most likely results and can often produce unwanted artifacts. This method is also known as Wiener–Kolmogorov prediction, after Norbert Wiener and Andrey Kolmogorov.

Both TP2 and Tecplot® have such capabilities. The blue mesh indicates the three-dimensional extent of the plume at the level of log(C)=-3. The white specks indicate were most of the particles will be seeded. Generating appropriate particle seeds will be covered in the next chapter.

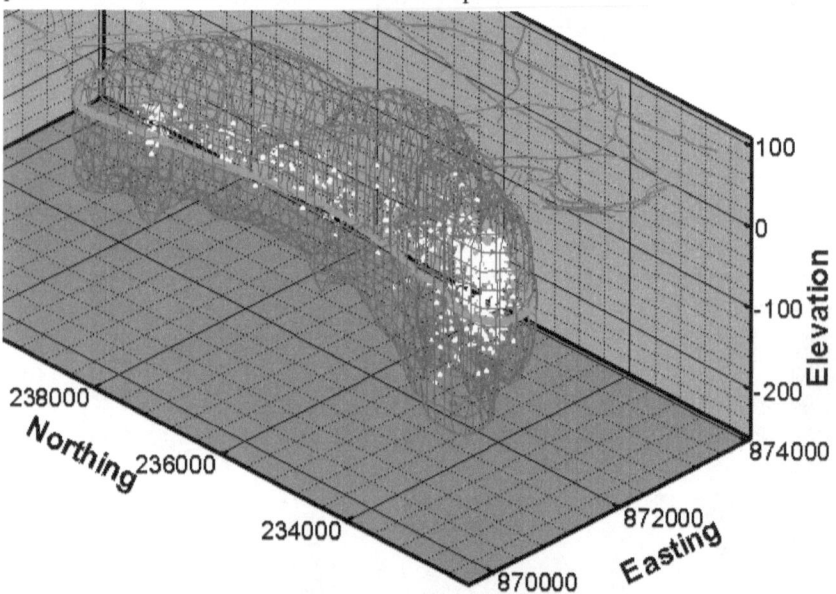

The shell at log(C)=-2 is shown in this next figure:

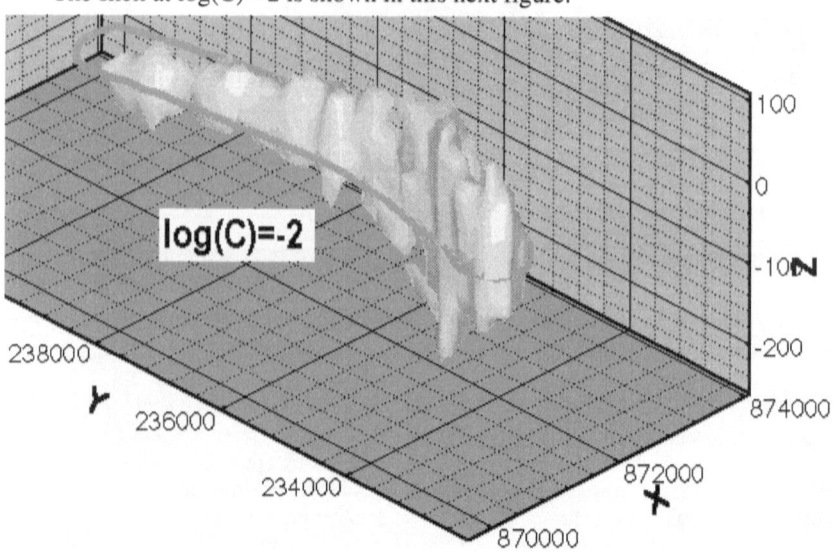

A horizontal slicing plane of concentrations is shown in this next figure:

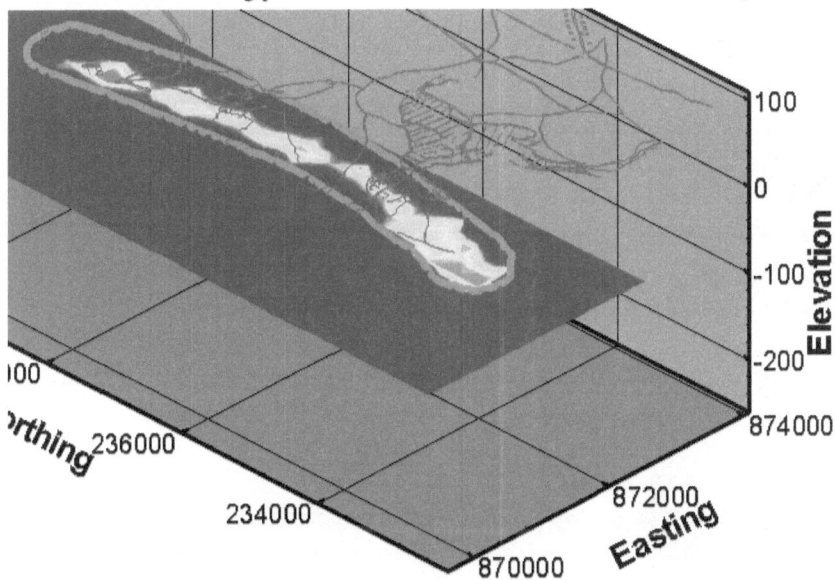

A vertical slicing plane of concentrations is shown in this next figure:

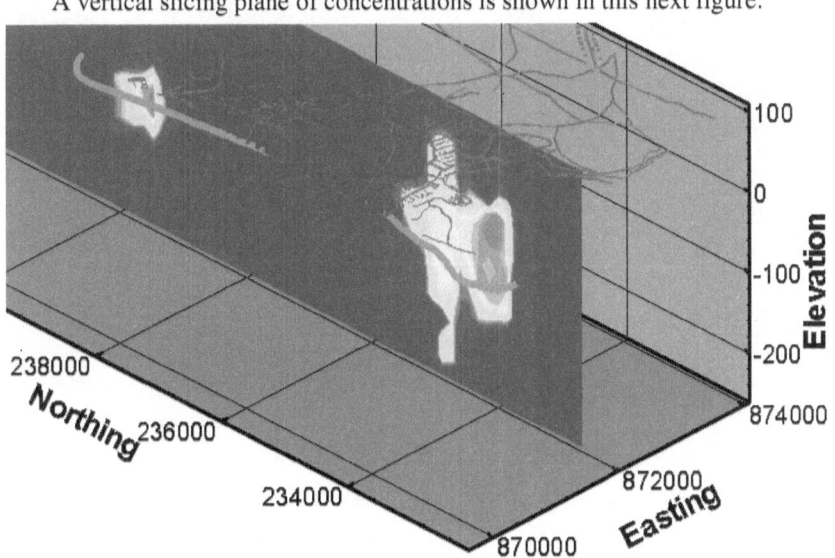

The most interestingly shaped plume we ever worked with was affectionately called the *starship*:

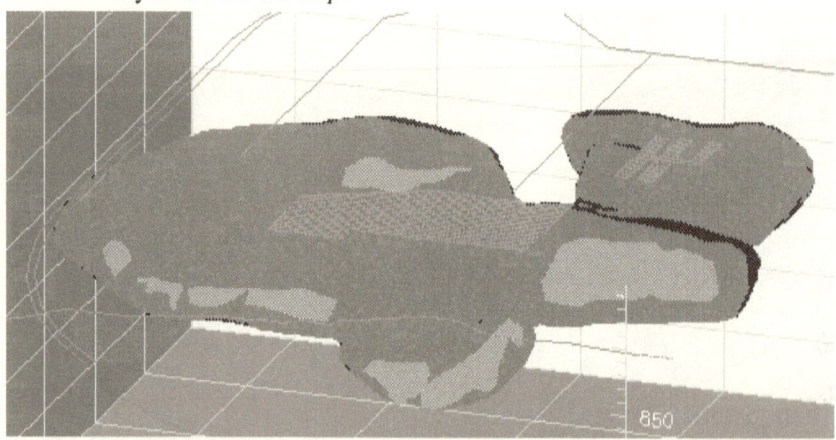

A view from the top:

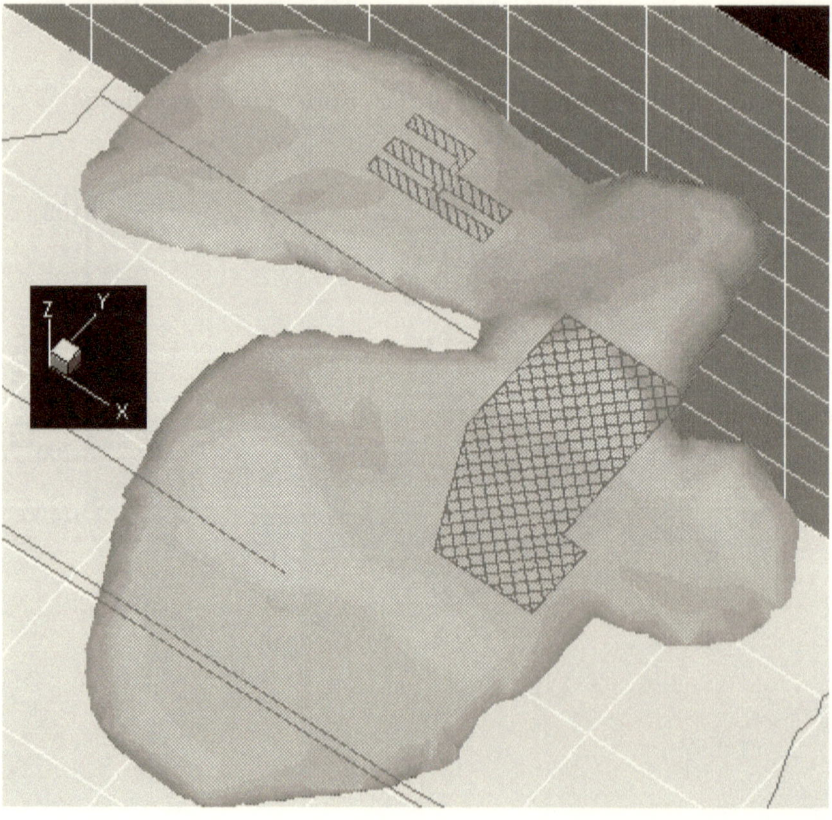

Chapter 9. Particle Seeds

Creating a *swarm* of particles that accurately represent an initial concentration field in three-dimensions is a complex task. PTRAX can handle particle seeds having different mass and/or concentration and can keep these separate, even coloring them differently, as shown in the following figure:

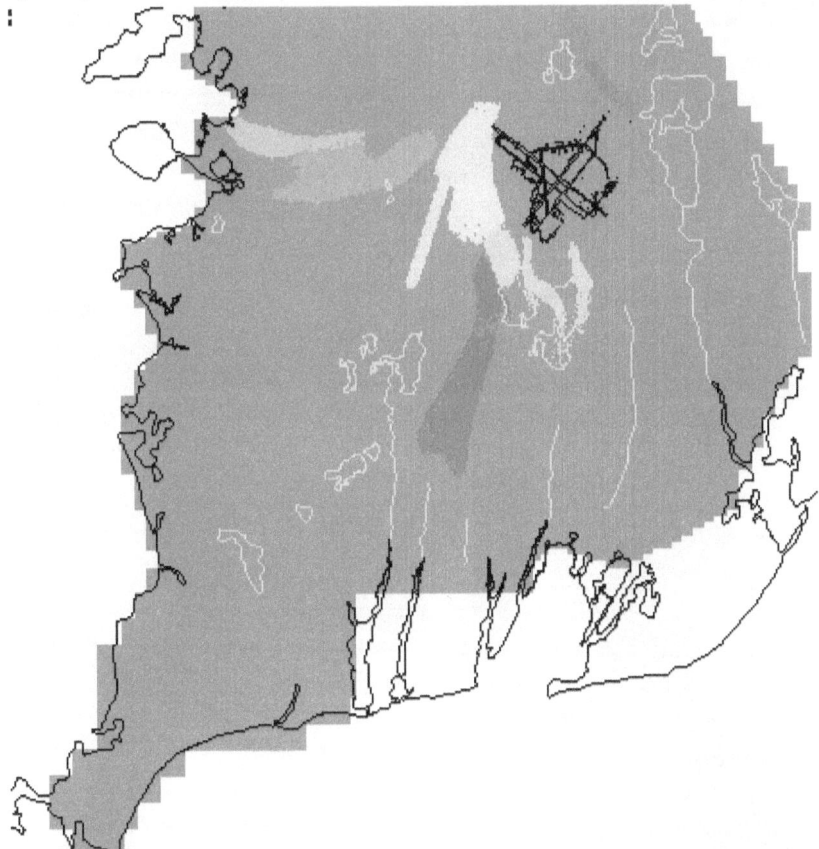

This is the first frame in an animation showing how the then current seven distinct contaminant plumes can be expected to evolve over time, should no steps be taken to remediate.

This next figure shows where wells were positioned to capture the contaminants. The circles are pump-and-treat wells and the plusses are monitoring wells to make sure the contaminant didn't get past the extraction wells. The calculated arrival of particles over time was also used to design the treatment systems, as different chemical substances and concentrations were expected.

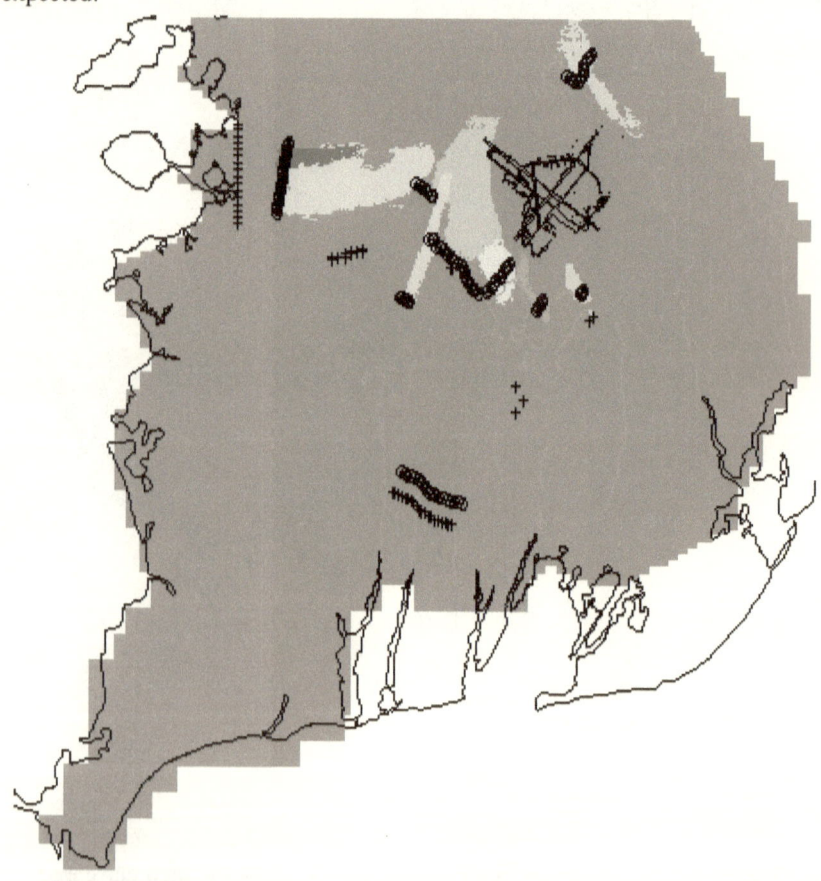

The first step in creating the particle seeds is defining the three-dimensional field describing the initial concentrations. This can be done by TP2 or Tecplot® resulting in a name.TB3 file. The format of this file is Nx, followed by the X values, Ny followed by the Y values, Nz followed by the Z values, and finally Nc=Nx*Ny*Nz followed by the concentrations. Both TP2 and Tecplot® can display 3D fields in several ways, including shells, as illustrated in the previous chapter, and slices along planes, as illustrated in this next figure:

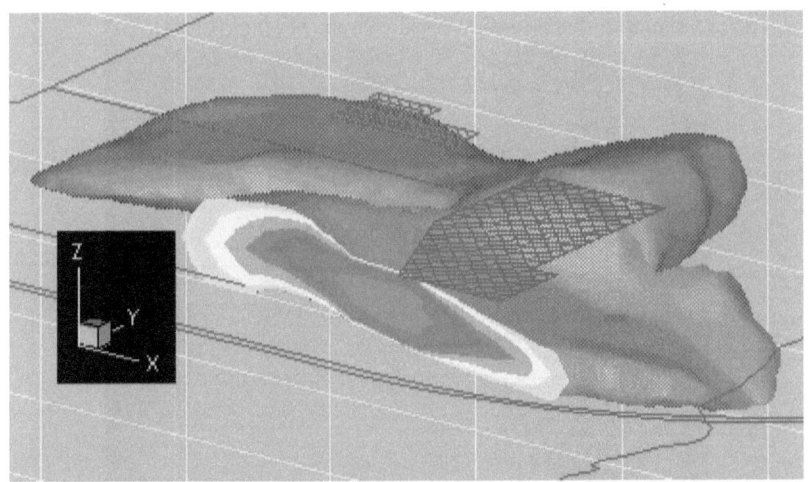

And a different-shaped plume:

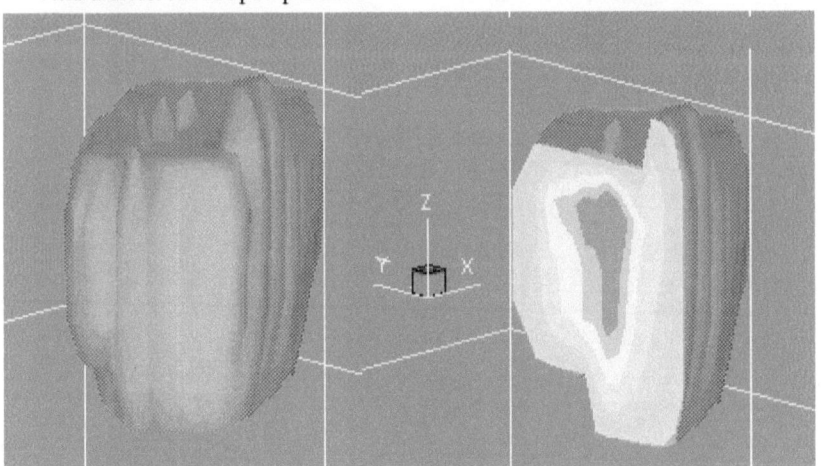

You will find 5 example three-dimensional concentration files in the online archive in folder examples\seeds. I have many more. These were all created with TP2 using inverse distance interpolation and also some smoothing, which is under the Tools menu.

The crosshairs in this next figure indicate the slicing planes in 3D:

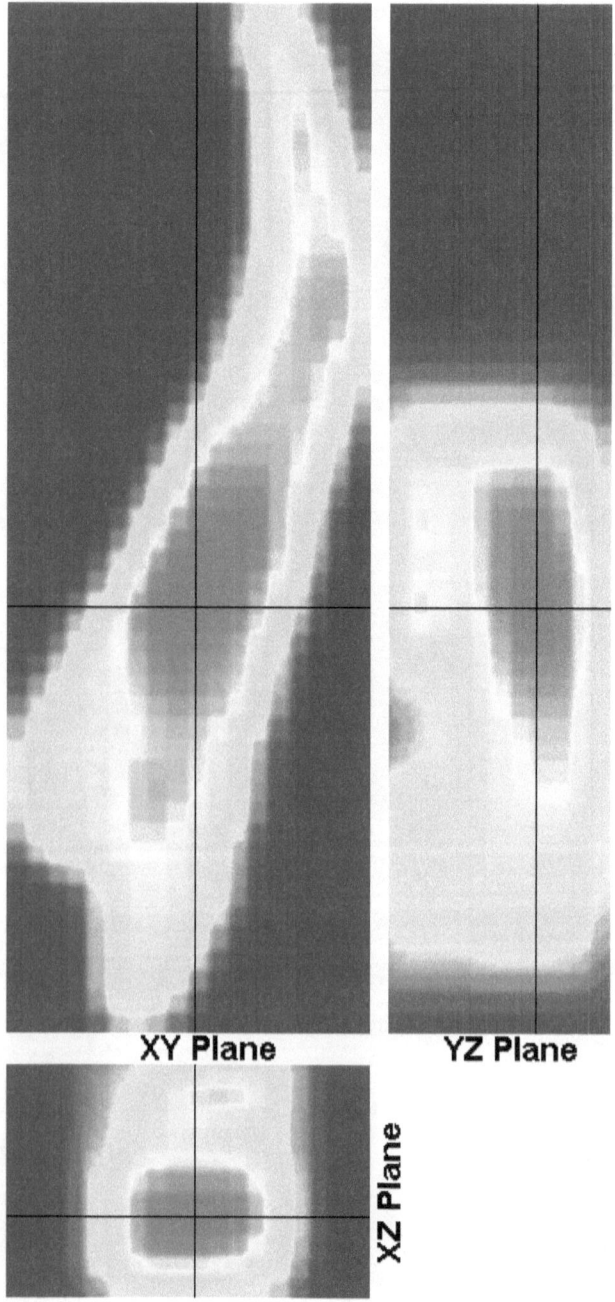

XY Plane

YZ Plane

XZ Plane

The seed generation program is seed3d.c, which outputs something like the following:

```
SEED3D/V1.13: create seed file from a .TB3 file
prefix: ash
seeds = 10000
extent of plume: ash.p2d
polygon: 222 points
concentration file: ash.TB3
seed file: ash.SED
field: 19x50x12=11400
concentration: -1.30103≤log(C)≤1.87935 µg/l
threshold = 10 µg/l
plan view area = 1221.94 acre (4.94502 km²)
plume volume = 31927.5 acre-ft (39380.8 Ml)
average thickness = 26.1285 ft (7.96395 m)
total plume mass = 915.705 kg
average concentration = 23.2526 µg/l
9679 actual seeds
total seed mass = 915.705 kg
TP2 command file: ash.TP2
seed mass: 3.23389E+006≤M≤3.77226E+006
```

It read the field from name.TB3 and the boundary from name.P2D and produces the seeds, which look something like:

```
* Note: mass is in æg-ft^3/liter, when
*        divided by volume in cubic feet,
*        this becomes æg/l or ppb.
*  X     Y      Z      Mass Time
857532 220962 -78.2251 3.37734E+006 0
857864 220970 -74.3877 3.37734E+006 0
857675 221093 -69.7951 3.37734E+006 0
857839 220809 -69.3116 3.37734E+006 0
857824 220942 -76.7429 3.37734E+006 0
857537 220776 -75.9333 3.37734E+006 0
857592 220805 -67.5764 3.37734E+006 0
857715 220787 -67.3602 3.41488E+006 0
857535 220889 -60.3033 3.41488E+006 0
857766 220977 -59.2923 3.41488E+006 0
857599 221001 -61.3863 3.41488E+006 0
857676 220762 -59.2853 3.41488E+006 0
857854 221056 -60.4611 3.41488E+006 0
857655 221085 -57.6919 3.41488E+006 0
857925 221499 -73.5912 3.48598E+006 0
857590 221316 -77.6624 3.48598E+006 0
857886 221217 -70.3733 3.48598E+006 0
857879 221527 -67.4268 3.48598E+006 0
857783 221289 -77.2488 3.48598E+006 0
857653 221465 -80.4957 3.48598E+006 0
857686 221170 -71.7453 3.48598E+006 0
```

The seeds are sprinkled throughout the domain, as shown in this next figure:

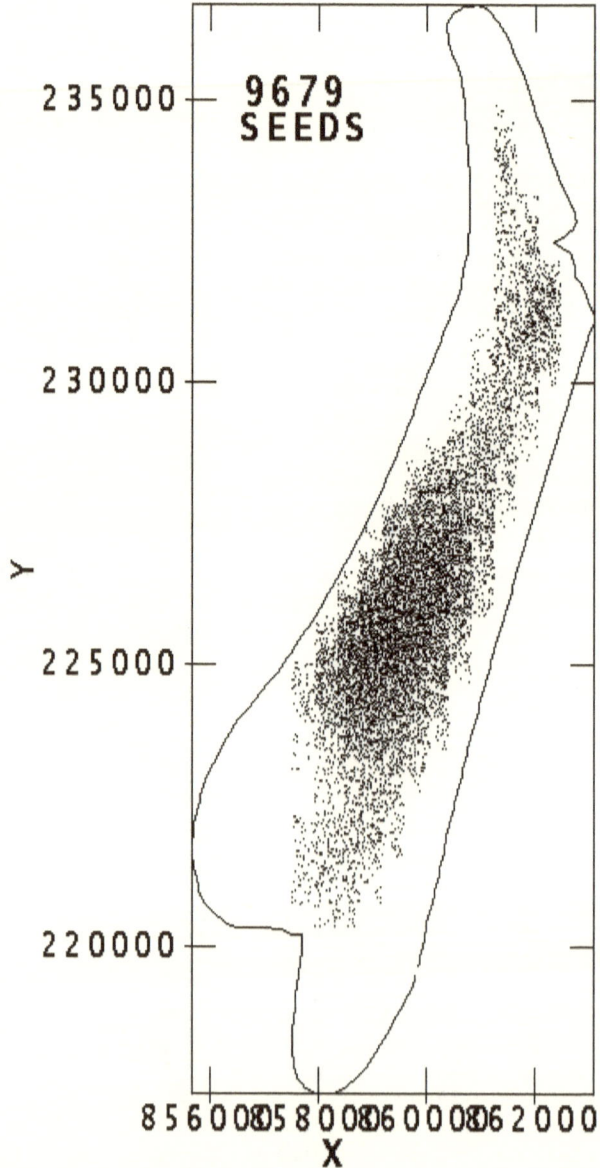

The actual number of seeds will be determined to most accurately represent the field. An approximate number of 10,000 was specified in this case. You can see from this figure that most of the seeds are placed in the highest concentration (i.e., red) zone from the previous slice graphic.

The core of seed3d.c is the following code:

```
for(x=1;x<Nx-1;x++)
  {
  X1=(Xn[x-1]+Xn[x])/2;
  X2=(Xn[x]+Xn[x+1])/2;
  dX=X2-X1;
  for(y=1;y<Ny-1;y++)
    {
    if(!InsidePolygon(Xp,Yp,Np,Xn[x],Yn[y]))
      continue;
    Y1=(Yn[y-1]+Yn[y])/2;
    Y2=(Yn[y]+Yn[y+1])/2;
    dY=Y2-Y1;
    for(z=1;z<Nz-1;z++)
      {
      Z1=(Zn[z-1]+Zn[z])/2;
      Z2=(Zn[z]+Zn[z+1])/2;
      dZ=Z2-Z1;
      dV=dX*dY*dZ;
      n=(Ny*z+y)*Nx+x;
      C=Cn[n];
      if(C<Cthr)
        continue;
      M=pow(10,C)*dV;
      m=max(1,(int)(ns*M/Mtot));
      Ms+=M;
      Ns+=m;
      M/=m;
      Mm=min(Mm,M);
      Mx=max(Mx,M);
      for(i=0;i<m;i++)
        {
        X=X1+(X2-X1)*drand();
        Y=Y1+(Y2-Y1)*drand();
        Z=Z1+(Z2-Z1)*drand();
        fprintf(fo,"%1G %1G %1G %1G 0\n",X,Y,Z,M);
        }
      }
    fprintf(stderr,"%li seeds\r",Ns);
    }
  }
```

Note that the TB3 file is assumed to contain the log of the concentrations. If not, you can modify the code accordingly. Also, the TB3 file is presumed to contain rectangular elements, though not necessarily the same length on each side, especially in the vertical.

Chapter 10. Animations

PTRAX automatically creates animations of the particle tracks. As this is a Windows® application, the GUI is continuously updated as the particles are tracked, which is per particle over time. What you really want is all of the particles over time. As the Hamiltonian approach tracks each particle sequentially, rather than simultaneously, as in the Lagrangian approach, the animations must also be created while the particles are being tracked and then written out after all of the calculations are complete. Compare the 1st and 24th frames (0 and 30 years) in this example shown previously:

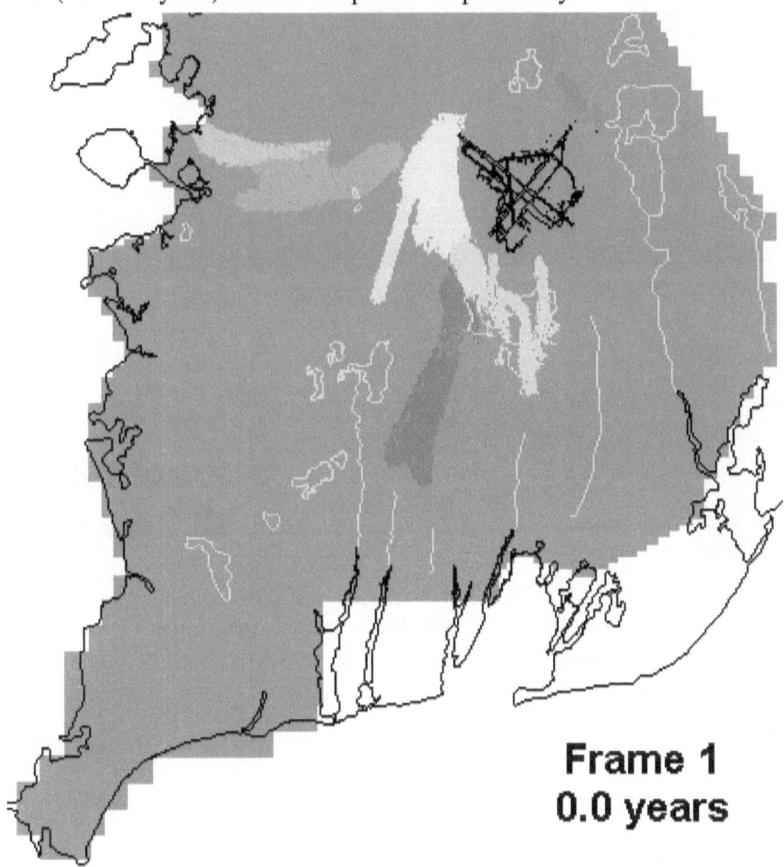

Frame 1
0.0 years

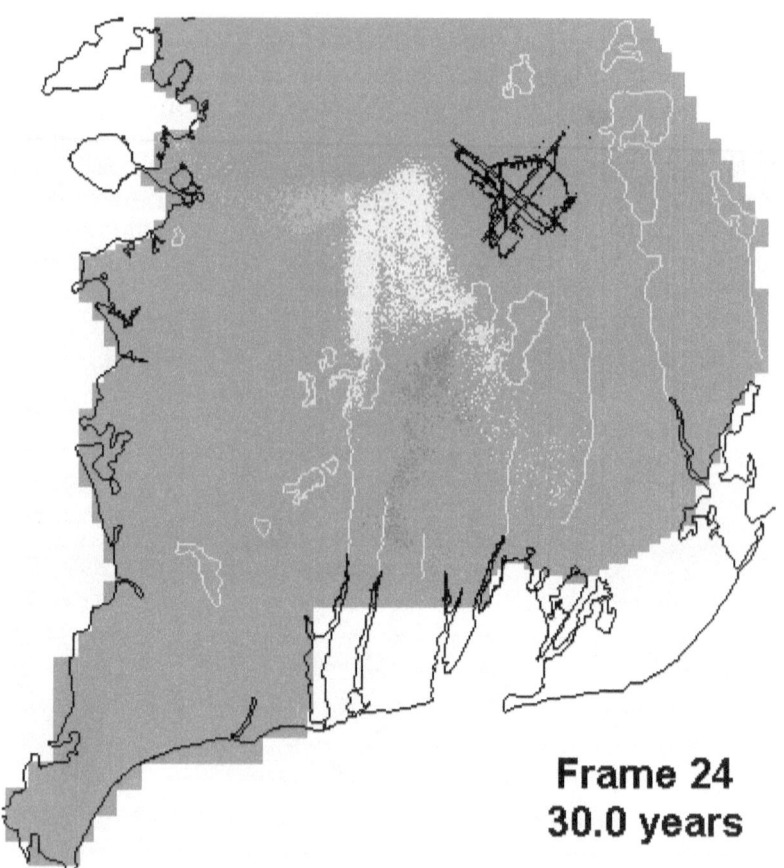

Frame 24
30.0 years

Notice that the particles have dispersed and also moved away from the original contamination zones. This particular site has the highest elevation in the center with groundwater flow outward, mostly to the west, south, and east, with very little flow to the north.

This next simulation achieved only partial capture of the contaminant. Compare the 1st and 20th frames (0 and 30 years).

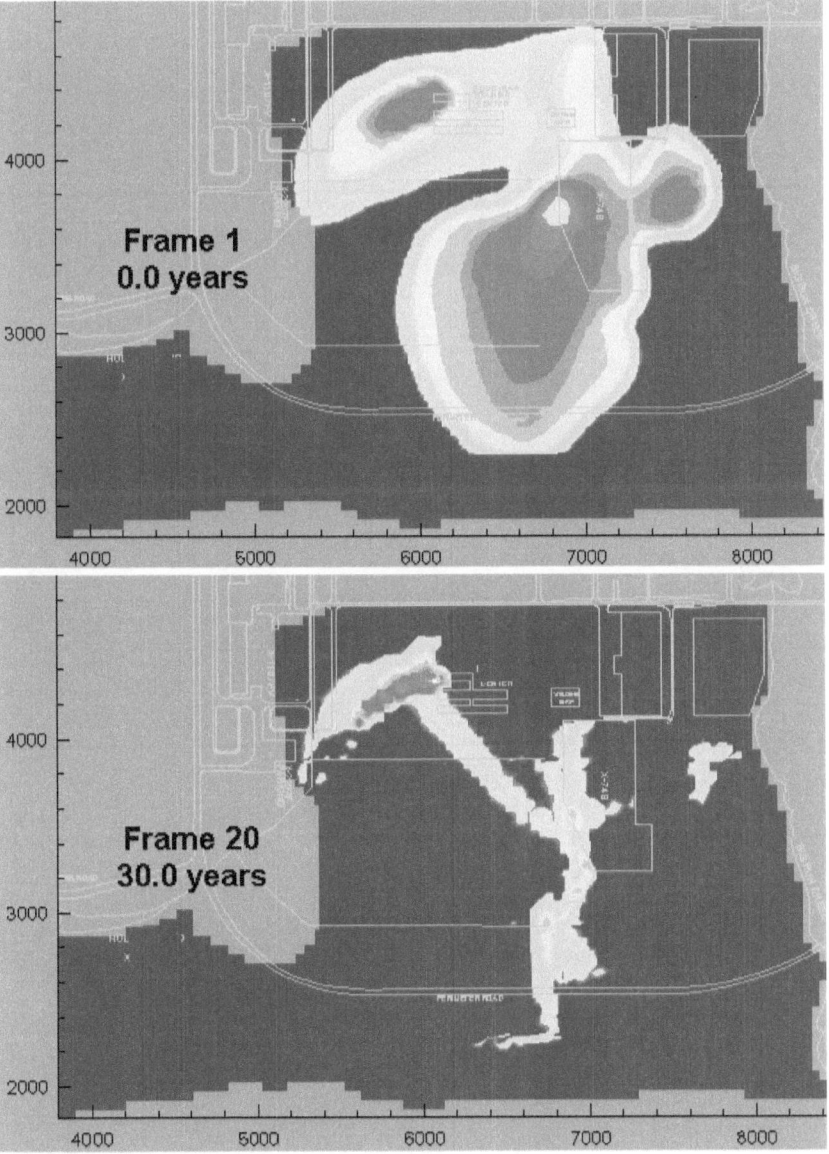

This is a good illustration of how modeling can shape remediation strategies. It's a lot cheaper to run a model twenty times than drill a bunch of wells and wait 30 years to find out what might or might not happen!

Chapter 11. Concentration Mappings

PTRAX automatically creates concentration snapshots at regular intervals as specified. These may or may not be set within model boundaries. Both TP2 and Tecplot® have the facility to overlay and embed images in fields to enhance the graphical presentation. The previous spaceship plume is shown in these next three figures, which are estimates of taking no remedial action.

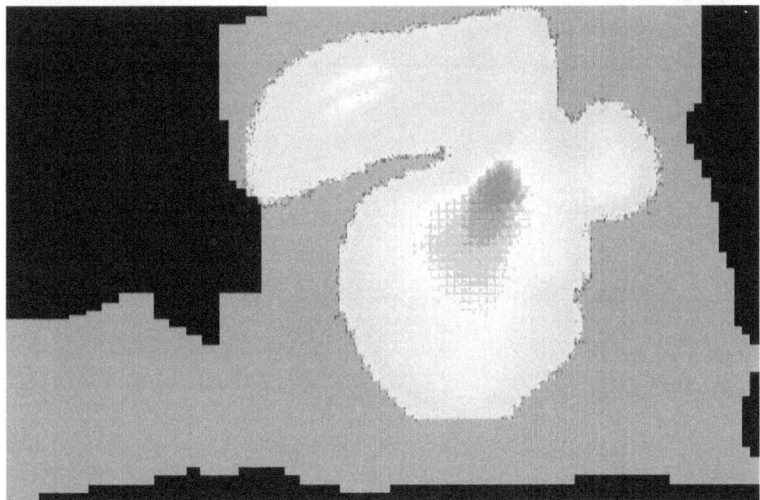

Frame 1 – 0.0 Years

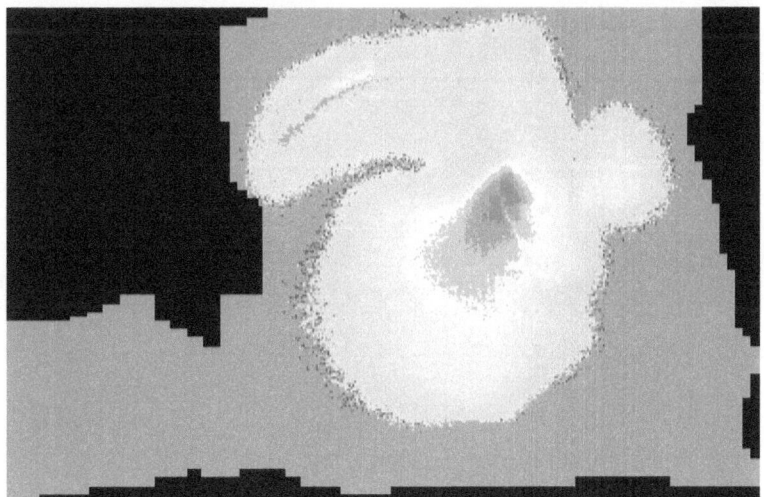

Frame 19 – 14.2 Years

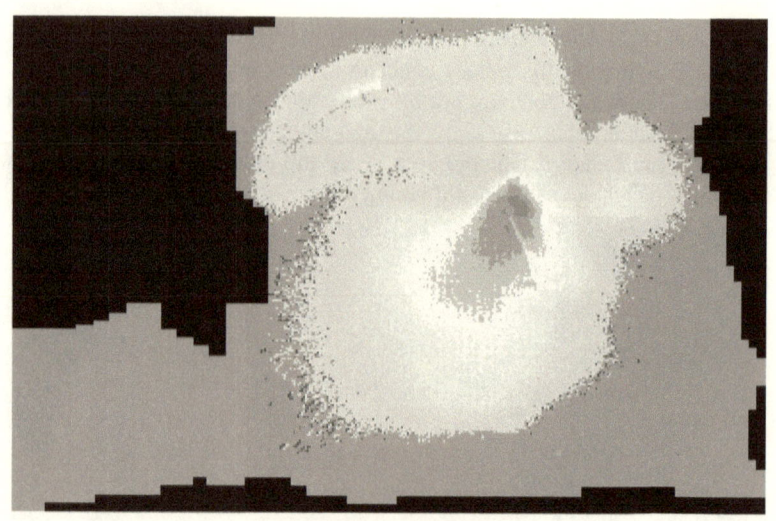

Frame 39 – 30 Years

PTRAX writes two- or three-dimensional table files (TB2 or TB3) for each concentration snapshot as well as images, such as these.

Chapter 12. Reverse Particle Tracking

Implementing reverse particle (i.e., back-) tracking is a real no-brainer. It's just a switch (BACK). When you read in the velocity field, simply reverse all the components. Adding this capability led us to one of the most interesting stories that I dare not share completely, considering the parties involved. Starting from what is and running backward in time to what might have started the current mess led is to find a vacant lot in a residential neighborhood where someone had apparently drained a tanker truck load of solvent that was supposed to be incinerated. So, build a model, make sure it agrees well with available data and then turn back the clock to see who left the mess and where!

The forward tracks for the example on page ii are:

Just add BACK to the command line or in the batch file (BEND.BAT), as in PTRAX BEND TRACK BACK WAIT OK, to track the same particles in reverse:

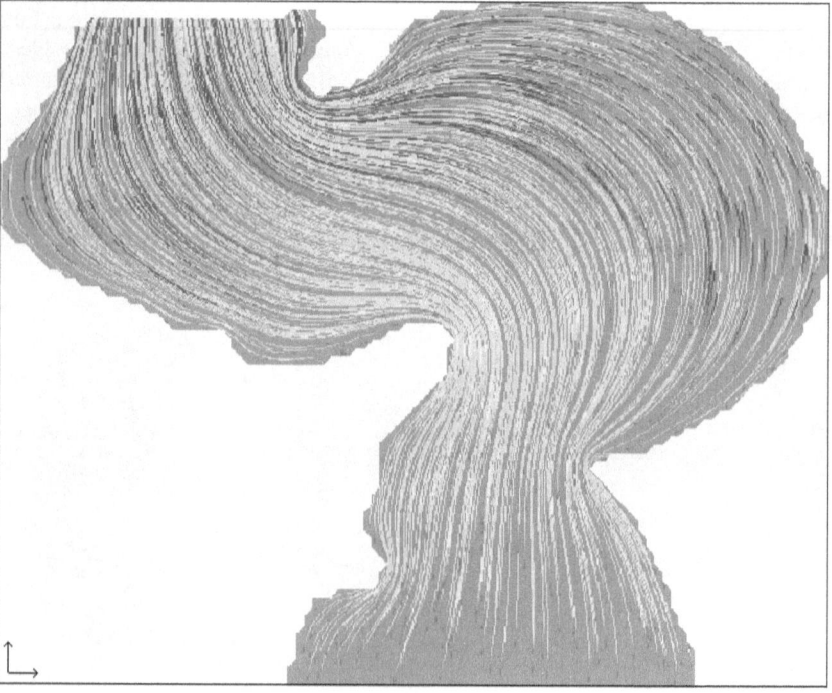

Chapter 13. Sources and Sinks

Wells can be used to inject and withdraw so that they are either sources or sinks. The flow model must be able to handle either. PTRAX can inject or remove particles. This is typical of such a system:

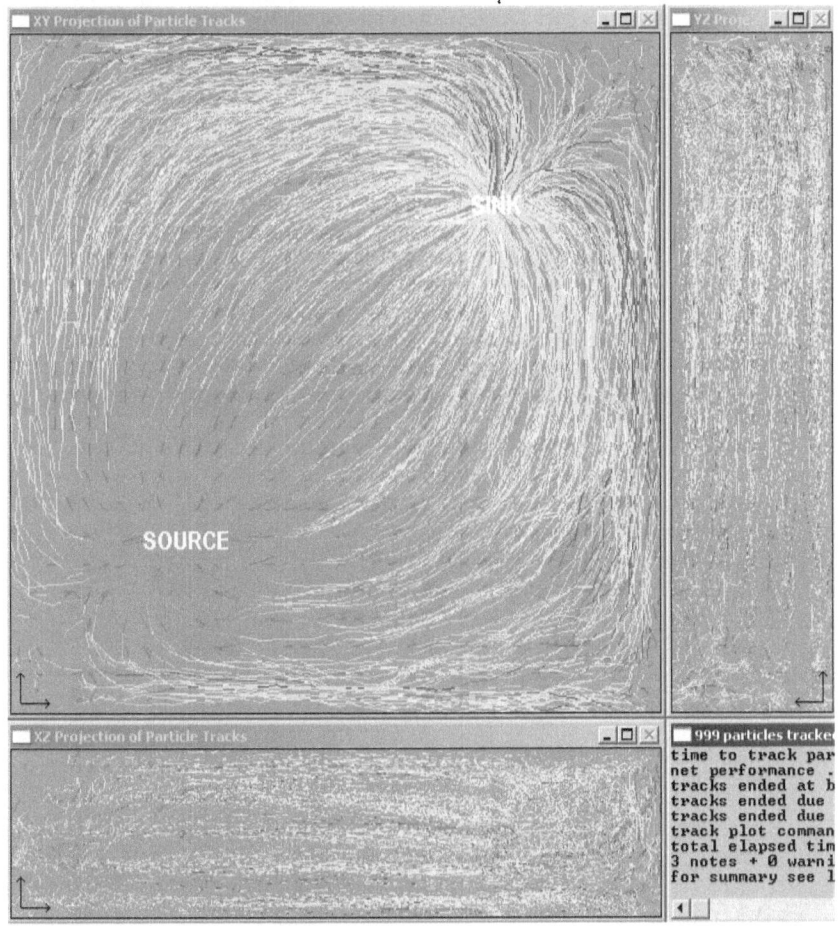

```
PTRAX/V5.01: Particle Tracker by Dudley J. Benton
animations:    enabled
fractures:     enabled
input format: Frac3D or ModFlow
parameter 2: TRACK
parameter 3: WAIT
parameter 4: OK
application prefix: SLAB
checking for existing output files
   none found
```

```
reading default parameters from file: SLAB.CFG
default parameters
  default seeds: 999
  track duration: 10950 days (30.0 years)
  time step: 1825 days (5.0 years)
  maximum steps along particle track: 999
  default porosity: 0.3
  default retardation factors: 1.1
  default dispersivities: 3,3,0.3 ft
  default matrix half-life: 1E+030 day(s)
  default fracture half-life: 1E+030 day(s)
  default velocities: 0.05,0,0 ft/day
  matrix diffusion coefficients: 0,0,0 ft²/day²
  synchronous time step default
  maximum times a particle can enter an element: 4
  maximum times a particle can enter a fracture: 4
  matrix tortuosity (0=none, 1=complete): 0
  Note: matrix tortuosity OFF
  fracture tortuosity (0=none, 1=complete): 0
  Note: fracture tortuosity OFF
reading MODFLOW files
Note: concentrations will NOT be divided by porosity
stray particles (outside domain) will be ignored
trapped particles will be ignored
include empty elements in snapshot files
create track file
MODFLOW basic input file: SLAB.BAS
  Title1: PREFIX: SLAB
  Title2: CREATED BY BUILD3D/V1.71
  grid: 37x25x5
  total cells: 4625
  active cells: 4625
  resulting nodes: 5928
  element type: HEXAHEDRA
MODFLOW block-centered flow file: SLAB.BCF
  06X6999.999
  06Y61000
  06Z6100
linking domain
  node:element links
  37000 links
  element:element links
  25280 internal faces
  2470 external faces (boundaries)
  2210 external elements
  2472 external nodes
  computing element volumes
grouping elements
  largest element: 27.027x40x20
```

```
22x14x2=616 groups
group size: 47.619x76.9231x100
sorting groups
indexing groups
there are 273 active and 343 empty groups
the smallest group is 135, containing 5 members
the largest group is 1, containing 20 members
the active groups contain an average of 17 members
MODFLOW binary flow file: SLAB.CBB
characteristic parameters
length threshold = 0.141774 feet
volume threshold = 9.99999E-005 ft^3
time threshold = 0.566894 day(s)
velocity threshold = 2.5009E-007 ft/day
mean velocity = 0.25009 ft/day
mean time to traverse element = 138.208 day(s)
synchronous time step = (automatic)
random walk
dispersion ON
diffusion OFF
scattered seeds: 999
particles tracked: 999
time to track particles: 1 seconds
average particles tracked per minute: 59940
average steps per particle: 27
average particle track: 425.621
average particle life: 1.12827E+006
average particle speed: 0.000377232
total   particle movement: Sp=425195
random particle movement: Rx/Sp=-0.00884701
random particle movement: Ry/Sp=-0.00415675
random particle movement: Rz/Sp=1.75661E-005
random time steps: 2681.7 ñ 395.558 day(s)
tracks ended at boundaries: 86
tracks ended due to circulation: 782
tracks ended due to maximum time: 131
```
Summary of Particle Tracking by Centroid of Mass

snap	year	mass	%mass	Xcentroid	Ycentroid	Zcentroid	Xradius	Yradius	Zradius
1	0	2.83E-05	100.0	512.5	518.3	50.8	282.9	280.7	28.3
2	5	2.77E-05	98.0	621.3	626.5	47.9	262.6	264.0	26.7
3	10	2.72E-05	96.1	672.9	673.0	46.3	232.3	241.4	25.7
4	15	2.66E-05	94.2	695.8	696.7	45.8	206.0	219.0	25.2
5	20	2.63E-05	92.9	709.9	711.8	45.4	187.4	194.7	24.9
6	25	2.61E-05	92.2	721.1	717.2	45.0	172.1	182.0	24.9
7	30	2.59E-05	91.5	727.3	720.2	44.7	160.2	170.8	24.9

```
plot command file: SLAB.TP2
Summary of Particle Tracking by Mass
```

snap	year	matrix	boundary	decay	trapped	pending	limbo
1	0	2.83E-05	0	0	0	0	0
2	5	2.77E-05	5.66E-07	0	0	0	0
3	10	2.72E-05	1.10E-06	0	0	0	0
4	15	2.66E-05	1.64E-06	0	0	0	0
5	20	2.63E-05	2.01E-06	0	0	0	0
6	25	2.61E-05	2.21E-06	0	0	0	0
7	30	2.59E-05	2.41E-06	0	0	0	0

Summary of Particle Tracking by Net Travel Distance

snap	year	%mass	dist	95%
1	0	100	1.77E-14	7.37E-14
2	5	97.998	196.354	231.939
3	10	96.0961	285.547	320.917
4	15	94.1942	326.068	356.601
5	20	92.8929	349.725	382.424
6	25	92.1922	365.715	407.887
7	30	91.4915	374.907	424.546

total elapsed time 3 seconds
3 notes + 0 warnings

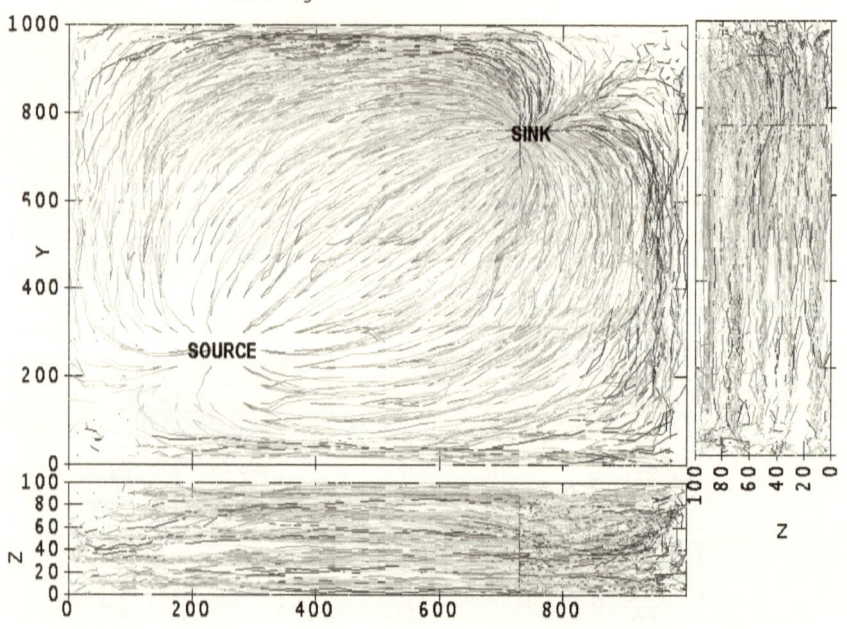

Chapter 14. Mosquito Tracking

Track mosquitoes? Not really... but sort of. One of my coworkers from the early days of PTRAX is Ken Black, a man of many talents. He volunteers with an international humanitarian organization that is concerned with one of the world's largest health crises: malaria in Zambia. This is a major concern of the World Health Organization. Even the Gates Foundation has funded various related projects, although not the one I'm about to describe.

Years ago Ken and I worked together on a project tracking runoff from surface water transporting contaminants. We began with GIS data, created 3D models, and modified several utilities (TP2 and PTRAX) to process the massive amounts of data. When the mosquito thing came up, Ken thought we might do something similar and so we did. The model is created on the fly from high-resolution topographic contours. It takes several hours to load and link three million elements and then track one million particles. The point is, where the particles end up are the most likely spots for stagnant water and mosquito breeding. Several of the areas identified by the model coincide with historical outbreaks of malaria. Here's an overview of the country:

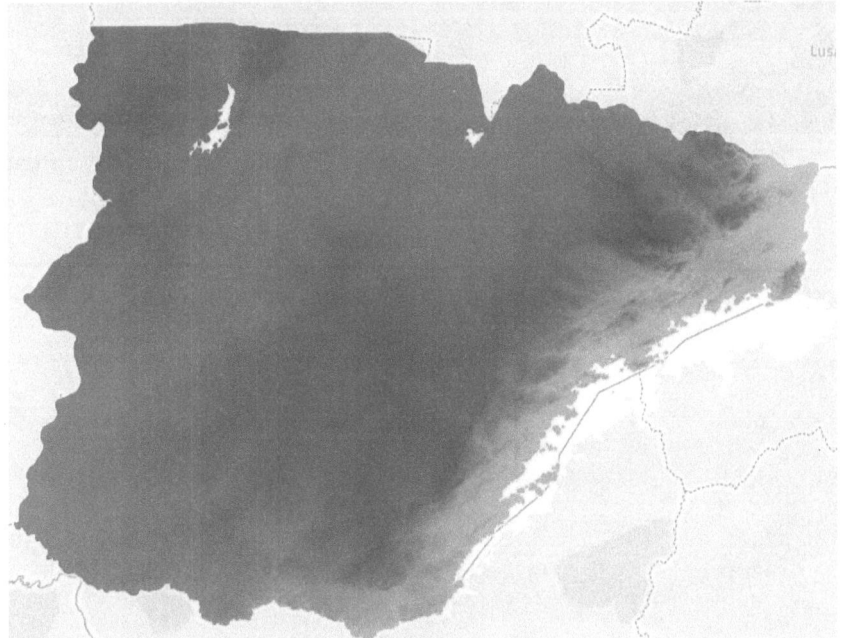

This figure shows the gradient of the surface elevation (TP2 will generate this automatically from the surface):

This figure shows the 3D velocity vectors (TP2 also generates these from the gradients):

This next figure shows the particle tracks and stagnant water sites:

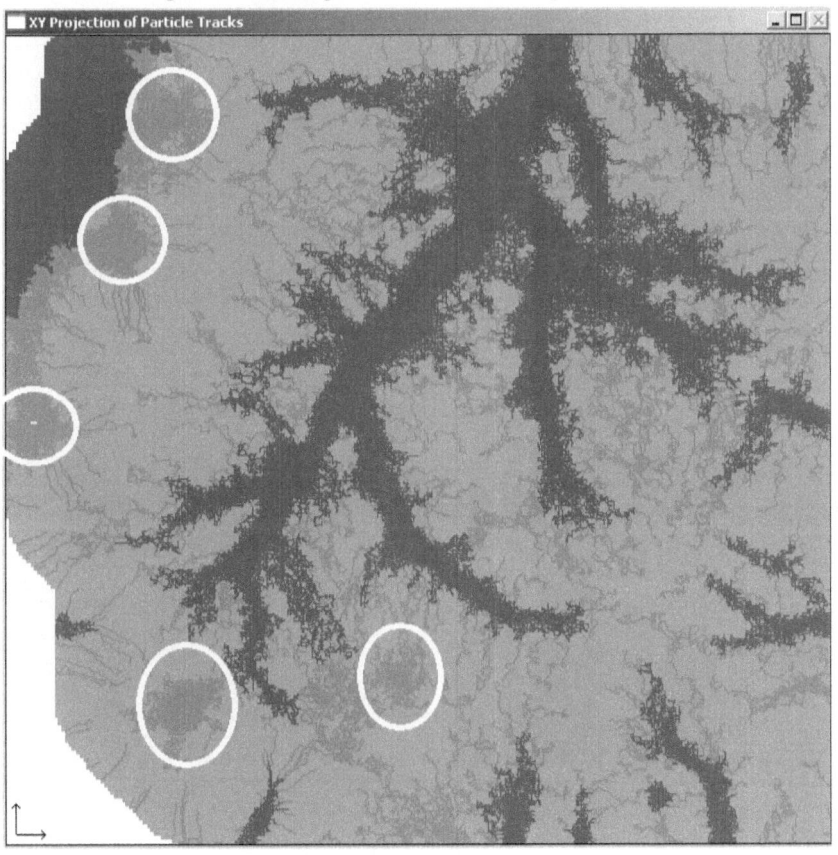

Another view of the particle tracks:

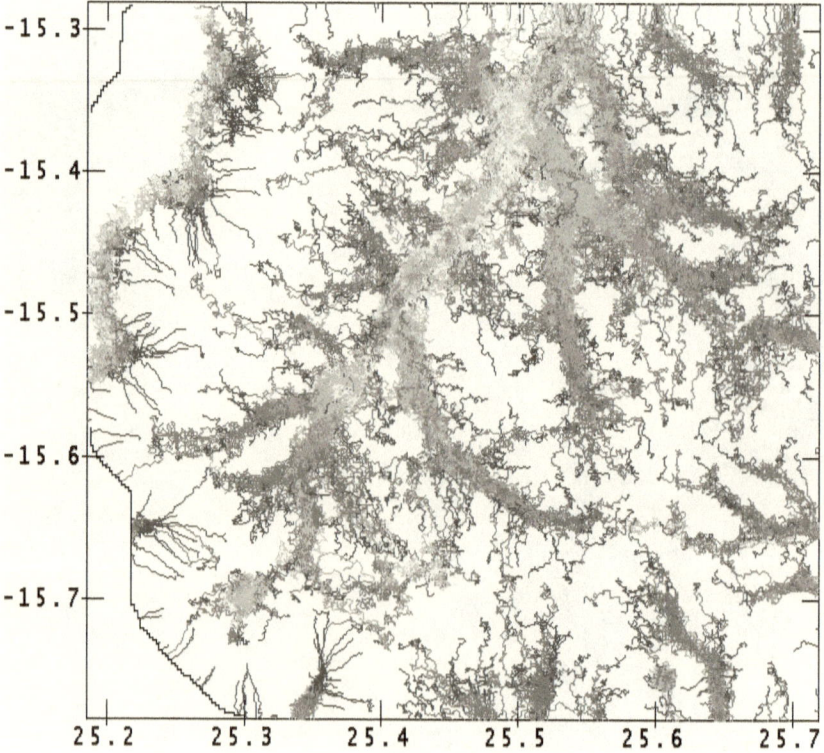

Chapter 15. Tracking Particles Inside Pipes

The purpose of this model was to simulate the injection of fluorescent dye into the intake of three pumps. The three pipes converge into a single larger one. There was only one dye injection pump and only one sampling device so that all three pumps could not be tested simultaneously. By injecting the dye three times and measuring the differences, the relative performance of the three pumps could be compared. The basic arrangement is:

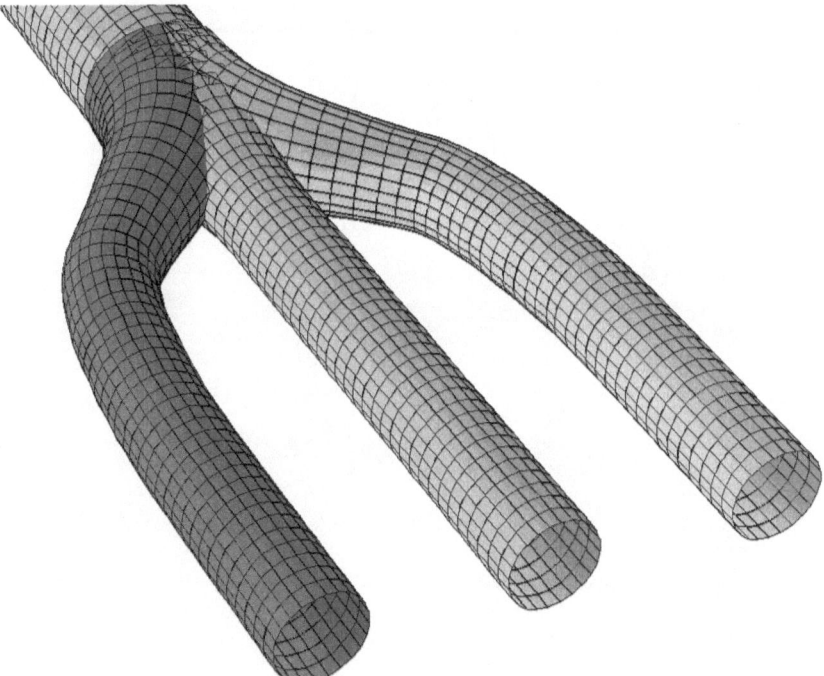

In order to get a significant statistical sample, 3,150,000 particles were found necessary, which took 14.0 minutes on a 3GHz Intel® processor.

```
PTRAX/V5.00avmf: Particle Tracker by Dudley J. Benton
compilation:    08/15/114:1734
animations:     enabled
fractures:      enabled
input format: Frac3D or ModFlow
available memory: 471518 KB
parameter 2: FOTO
parameter 3: WALL
parameter 4: OK
parameter 5: WAIT
application prefix: BLOCKS
file=BLOCKS.NDE    date=08/15/114:1824
file=BLOCKS.ELM    date=08/14/114:1909
```

```
file=BLOCKS.VEP    date=08/16/114:0104
file=BLOCKS.SED    date=09/10/119:2315
file=BLOCKS.WAL    date=08/14/114:2236
checking for existing output files
  none found
reading default parameters from file: BLOCKS.CFG
default parameters
  default seeds: 999
  track duration: 450 days (1.2 years)
  time step: 5 days (0.0 years)
  maximum steps along particle track: 999
  default porosity: 1
  default retardation factors: 1.1
  default dispersivities: 0,0,0 ft
  default matrix half-life: 1E+030 day(s)
  default fracture half-life: 1E+030 day(s)
  default velocities: 0,0,0 ft/day
  matrix diffusion coefficients: 0.0001,0.0001,0.0001
    ft²/day²
  synchronous time step default
  maximum times a particle can enter an element: 9
  maximum times a particle can enter a fracture: 9
  matrix tortuosity (0=none, 1=complete): 0
  Note: matrix tortuosity OFF
  fracture tortuosity (0=none, 1=complete): 0
  Note: fracture tortuosity OFF
reading FRAC3D files
Note: concentrations will NOT be divided by porosity
stray particles (outside domain) will be ignored
trapped particles will be ignored
include empty elements in snapshot files
create wall history file
FRAC3D node file: BLOCKS.NDE
  36 nodes
  -186X618
  06Y6320
  -56Z65
FRAC3D element file: BLOCKS.ELM
  8 elements
  element type: HEXAHEDRA
linking domain
  node:element links
  64 links
  element:element links
  14 internal faces
  34 external faces (boundaries)
  8 external elements
  36 external nodes
  computing element volumes
```

```
  Note: elements with clockwise orientation: 6
grouping elements
  largest element: 15x160x10
  2x2x2=8 groups
  group size: 36x320x10
  sorting groups
  indexing groups
  there are 1 active and 7 empty groups
  the smallest group is 1, containing 8 members
  the largest group is 1, containing 8 members
  the active groups contain an average of 8 members
FRAC3D velocity file: BLOCKS.VEP
  velocities: 8 (element-based)
using default properties
using default transport properties
characteristic parameters
  length threshold = 0.0322174 feet
  volume threshold = 1.152E-007 ft^3
  time threshold = 0.00644348 day(s)
  velocity threshold = 5E-006 ft/day
  mean velocity = 5 ft/day
  mean time to traverse element = 3.71769 day(s)
  synchronous time step = (automatic)
random walk
  dispersion OFF
  diffusion ON
wall file: BLOCKS.WAL
walls have the following dimensions
wall          area
  1            200
  2            200
wall:element links 2
seed file: BLOCKS.SED  3150000 seeds
  particles tracked: 3150000
  time to track particles: 13.9 minutes
  average particles tracked per minute: 226001
  average steps per particle: 51
  average particle track: 256.494
  average particle life: 176.015
  average particle speed: 1.45723
  total   particle movement: Sp=8.07957E+008
  random particle movement: Rx/Sp=-4.71776E-007
  random particle movement: Ry/Sp=-3.49909E-006
  random particle movement: Rz/Sp=3.60261E-007
  random time steps: 0.682491 ñ 0.022903 day(s)
  tracks ended at boundaries: 1194416
  tracks ended at capture walls: 1945092
  tracks ended due to circulation: 10492
Summary of Particle Tracking by Centroid of Mass
```

```
sorting wall histories
  wall capture file: BLOCKS.CAW
Summary of Particle Tracking by Mass
Summary of Particle Tracking by Net Travel Distance
snap  year %mass       dist       ñ95%
available memory: 401242 KB
total elapsed time 14.0 minutes
```

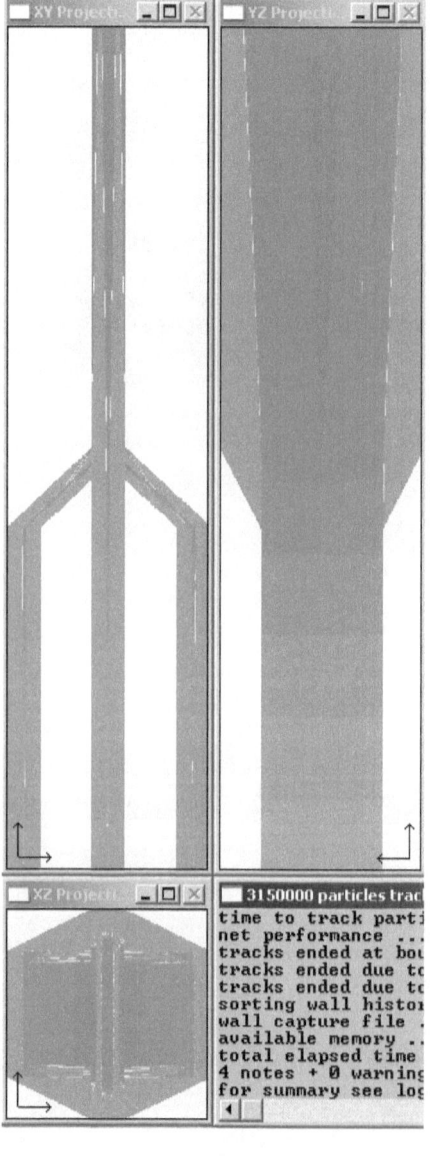

The analytical solution is:

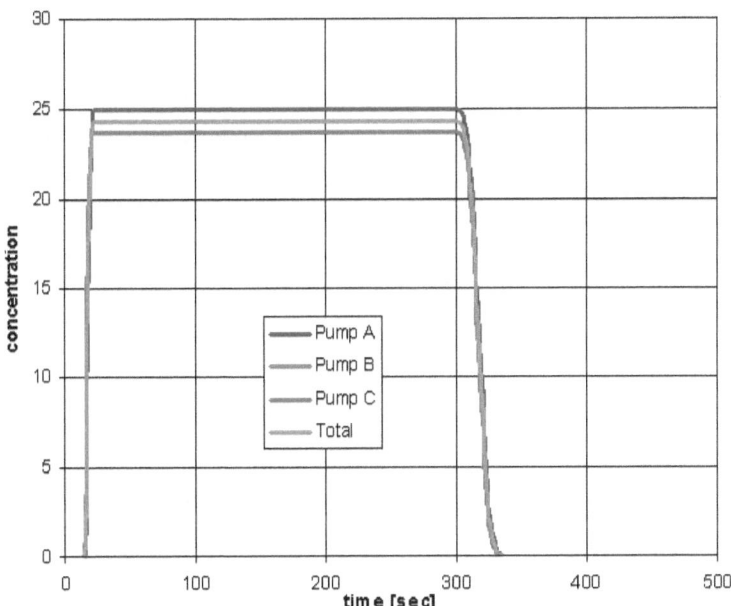

The simulated solution based on tracking three million particles:

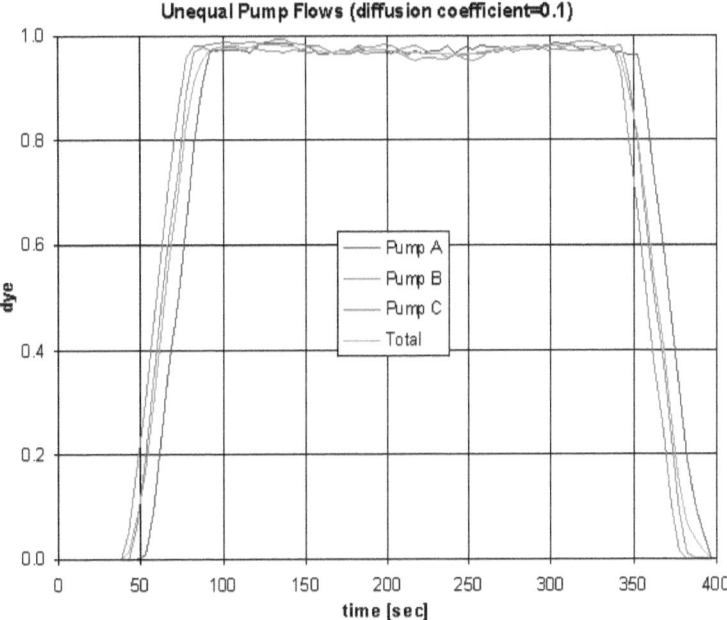

The particle tracking solution is more realistic and closer to the measured results, including delayed rise on the left and the spread between the blue and green curves on the right side between 350 and 400 seconds. The analytical solution is amorphous (smeared), while the particles take a finite amount of time to disperse through the media. The velocities were scaled so that seconds became years, which was easier than modifying the code to accept the shorter time units.

Appendix A. Potential Fields

Potential fields satisfy Laplace's partial differential equation. In Cartesian coordinates this is:

$$\frac{\partial^2 \varphi}{dx^2} + \frac{\partial^2 \varphi}{dy^2} + \frac{\partial^2 \varphi}{dz^2} = 0 \tag{A.1}$$

where φ is the potential. In cylindrical coordinates, this can be written:

$$\frac{1}{r}\frac{\partial}{\partial r}\left(r\frac{\partial \varphi}{\partial r}\right) + \frac{1}{r^2}\frac{\partial^2 \varphi}{\partial \theta^2} + \frac{\partial^2 \varphi}{dz^2} = 0 \tag{A.2}$$

Inviscid flow (i.e., flow of a fluid with zero or negligible viscosity or having an effectively infinite Reynolds number) satisfies this partial differential equation. Electrical and magnetic fields can often be described by this same partial differential equation. Heat conduction and molecular diffusion when linear in nature can be described by this same partial differential equation. This means that analytical solutions developed for these otherwise diverse disciplines are equally applicable to any other potential field.

Two-dimensional potential solutions are often expressed using complex variables (x+iy), where i=√-1. These complex solutions can be surprisingly simple formulas. Several examples can be found in the online archive in folder examples\potflow. Many more examples can be found in the online archive accompanying my text, *Complex Variables*, which can be freely downloaded from the same site.

For example, a source in polar coordinates is simply:

$$w = Q(\ln r + i\theta) \tag{A.3}$$

An irrotational (i.e., free) vortex is:

$$w = \frac{Q}{2\pi}(\theta - i\ln r) \tag{A.4}$$

Flow over a corner, step, or wedge is:

$$w = z^n \tag{A.5}$$

Stream Function

In two-dimensional flow, we can often—though not always—derive a stream function, whose curves of constant value are called streamlines. Streamlines are the paths that a particle would take if traveling through the field. The stream function is most often given the symbol ψ (Greek psi). The stream function is perpendicular to the potential function. In Cartesian coordinates this relationship can be expressed:

$$\frac{\partial \psi}{\partial x} = -\frac{\partial \phi}{\partial y}$$
$$\frac{\partial \psi}{\partial y} = \frac{\partial \phi}{\partial x}$$

(A.6)

It can be shown through the chain rule and combining A.6 with A.1 that the stream function also satisfies Laplace's Equation. Solutions exist (especially in 3D) where there is a stream function but no velocity potential and vice versa.

Irrotational Flow

When reading about potential flow, you may come across a statement that it is *irrotational*. You may also come across an example (such as in potflow.c and track1.c) that mentions *circulation* (as in flow over a cylinder with circulation, see page 5). How can this be? It depends on what you mean by *irrotational* and *circulation*. These have very specific mathematical definitions. To say that a flow (or any field) is *irrotational*, is to say that the *curl* of the velocity vector (i.e., the *vorticity*, which is also a vector) is zero. This is written:

$$\vec{\omega} = \nabla \times \vec{V} = 0$$

(A.7)

where ∇ is the del operator and V is the vector velocity:

$$\vec{V} = u\hat{i} + v\hat{j} + w\hat{k}$$

(A.8)

In Cartesian coordinates, the curl of V is:

$$\nabla \times \vec{V} = \left(\frac{\partial w}{\partial y} - \frac{\partial v}{\partial z} \right)\hat{i} + \left(\frac{\partial u}{\partial z} - \frac{\partial w}{\partial x} \right)\hat{j} + \left(\frac{\partial v}{\partial x} - \frac{\partial u}{\partial y} \right)\hat{k}$$

(A.9)

Appendix B. Boundary Element Method

The boundary element method arises from Green's Lemma[7]. This remarkable relationship between the integral of a function over an area and the integral of a corresponding function around the perimeter of the same domain is usually covered in advanced calculus. Green's Lemma can be expressed by the following integral:

$$\iint \nabla^2 \varphi \, dA = \int \frac{\partial \varphi}{\partial n} \, dS \tag{B.1}$$

In Equation B.1 φ is the potential, dA is the differential area within the domain, dS is a differential distance along the boundary, and n is the normal (perpendicular) at each location along the boundary. The potential could be simply that (i.e., electrostatic potential or invicid flow), temperature, concentration, stress, strain, or anything else that satisfies Laplace's equation.

We must also have a *fundamental* solution to Laplace's equation—in this case, a general, homogeneous solution (i.e., works for any case and is zero on the right-hand side). The derivation is a bit circuitous, except in polar coordinates.

$$\varphi = \ln\left(\frac{1}{r}\right) \tag{B.2}$$

We can at least show by substitution that Equation B.2 satisfies Laplace's equation in polar coordinates:

$$\frac{1}{r}\frac{\partial}{\partial r}\left(r\frac{\partial \varphi}{\partial r}\right) + \frac{1}{r^2}\frac{\partial^2 \varphi}{\partial \theta^2} = 0 \tag{B.3}$$

For Cartesian (x,y) coordinates, r in the above equation is the distance from some as yet unspecified location $(x=a, y=b)$, or $r^2=(x-a)^2+(y-b)^2$. We further propose that the potential, φ, and the derivative with respect to the normal, $\partial\varphi/\partial n$, have some specific, finite, non-trivial value at each of the points along the boundary. Consider two points along the boundary (1 and 2). We can write Equation B.2 for these two points:

---

[7] George Green (1729-1841): British mathematical physicist best known for work with electric fields and magnetism.

$$\varphi(r) = \left\{ \frac{\varphi_1 (r - r_2)}{\ln\left(\dfrac{1}{r_1}\right)} - \frac{\varphi_2 (r - r_1)}{\ln\left(\dfrac{1}{r_2}\right)} \right\} \frac{\ln\left(\dfrac{1}{r}\right)}{(r_1 - r_2)} \tag{B.4}$$

At x_1,y_1 $r=r_1$ and at x_2,y_2 $r=r_2$ so that $\varphi(r_1)= \varphi_1$ and $\varphi(r_2)= \varphi_2$; therefore, Equation B.4 is the particular form of the fundamental equation that satisfies Laplace's equation and matches at these two points along the boundary. We can construct a similar equation for the right side of Equation B.1. When integrate Equation B.4 from point 1 to point 2, we will get some constant times φ_1 plus some other constant times φ_2. We will use $H_{i,j}$ to represent these constants on the left side and $G_{i,j}$ to represent the corresponding constants on the right side of Equation B.1. The indices i and j indicate each segment along the boundary and each pair corresponding to each segment. The resulting set of equations can be written:

$$\sum_{i=}^{n}\sum_{j=1}^{n} H_{i,j}\varphi_j = \sum_{i=}^{n}\sum_{j=1}^{n} G_{i,j}\frac{\partial \varphi_j}{\partial n} \tag{B.5}$$

Equation B.5 constitutes a set of simultaneous linear equations for the potential function and its derivative at the n points along the boundary. We will need three different formulas for the integrals. We can readily integrate this equation around the boundary except at the two points (here, 1 and 2). At those points $r=r_1$ or $r=r_2$ and the standard result is indeterminate. We use a different formula for that one segment. We use these two formulas for points along the boundary. For points inside the boundary we use a third formula. We need this third integral to evaluate the results of the solution inside the boundary.

I have been vague up until this point on exactly what formulas are integrated and how, sparing you the gory details. In most applications, numerical integration (e.g., Gauss Quadrature) is used, because the analytical integral is unknown to the programmer. In fact, that's the way I present this as an example in my book *Numerical Calculus*. Here, you have the benefit of the analytical solution, which is precise, instead of a numerical solution that is approximate. Of course, I didn't figure this out the hard way. I used Maple® to do that for me. The resulting formulas are indeed tedious. You will find everything you need (source code and examples) in the online archive in folder examples\bem.

Appendix C. Explicit Runge-Kutta Methods

Runge-Kutta is a type of marching method in that we start with some initial values and then step along through time (or space). We will first consider explicit methods. The seminal reference on the Runge-Kutta and related methods was published by Butcher.[8] There are countless articles on the Web dealing with Runge-Kutta. In marching methods, we consider differential equations of the following form:

$$\frac{dy}{dx} = f(x, y(x)) \tag{C.1}$$

As we shall see, higher order differentials are easily handled by extending this formula. The initial position is represented by x and the time step, Δx, is represented by h. The symbol k is used to represent some particular value of $f(x,y(x))$. The simplest procedure is known as Euler's explicit method, which is implemented:

$$k_1 = f(x, y(x))$$
$$y(x+h) = y(x) + hk_1 \tag{C.2}$$

This is exactly the same as:

$$y_{x+\Delta x} = y_x + \Delta x \left.\frac{dy}{dx}\right)_x \tag{C.3}$$

Euler's explicit method is sometimes called 1st order Runge-Kutta. In general, these and similar methods can be expressed by the following formula, where n is the number of steps, which is not necessarily the same as the order:

$$k_1 = f(x, y)$$
$$k_2 = f(x + c_2 h, y + h(a_{21}k_1))$$
$$k_3 = f(x + c_3 h, y + h(a_{31}k_1 + a_{32}k_2)) \tag{C.4a}$$

$$\cdots$$

$$k_i = f\left(x + c_i h, y + h\sum_{j=1}^{i-1} a_{ij}k_j\right) \tag{C.4b}$$

$$y = y + h\sum_{i=1}^{n} b_i k$$

---

[8] Butcher, J. C., The Numerical Analysis of Ordinary Differential Equations: Runge-Kutta and General Linear Methods, John Wiley & Sons Ltd., New York, 1987.

Butcher Tableaus

Butcher expressed the preceding set of equations in tabular form, called a tableau, having the following form:

$$
\begin{array}{c|cccccc}
c_1 & a_{11} & a_{12} & a_{13} & \cdots & a_{1n} \\
c_2 & a_{21} & a_{22} & a_{23} & \cdots & a_{2n} \\
\cdots & \cdots & \cdots & \cdots & \cdots & \cdots \\
c_n & a_{n1} & a_{n2} & a_{n3} & \cdots & a_{nn} \\
\hline
 & b_1 & b_2 & b_3 & \cdots & b_n
\end{array}
\qquad (C.5)
$$

The Butcher tableau for Euler's explicit method (Equation 2.2) is:

$$
\begin{array}{c|c}
0 & 0 \\
\hline
 & 1
\end{array}
\qquad (C.6)
$$

We will present all of these methods in this way and then implement them in a code that can handle any formula in this form. There are three common variants of 2nd order Runge-Kutta. The first variant is:

$$
\begin{array}{c|cc}
0 & 0 & 0 \\
1/2 & 1/2 & 0 \\
\hline
 & 0 & 1
\end{array}
\qquad (C.7)
$$

The second variant is called Huen's method:

$$
\begin{array}{c|cc}
0 & 0 & 0 \\
1 & 1 & 0 \\
\hline
 & 1/2 & 1/2
\end{array}
\qquad (C.8)
$$

The third variant is called Ralston's method:

$$
\begin{array}{c|cc}
0 & 0 & 0 \\
2/3 & 2/3 & 0 \\
\hline
 & 1/4 & 3/4
\end{array}
\qquad (C.9)
$$

There are also two common variants of 3rd order Runge-Kutta. The first is:

0	0	0	0
1/2	1/2	0	0
1	-1	2	0
	1/6	2/3	1/6

$$(C.10)$$

The second variant of 3rd order Runge-Kutta is:

0	0	0	0
1/3	1/3	0	0
2/3	0	2/3	0
	1/4	0	3/4

$$(C.11)$$

There are also two common variants of 4th order Runge-Kutta. The first is:

0	0	0	0	0
1/2	1/2	0	0	0
1/2	0	1/2	0	0
1	0	0	1	0
	1/6	1/3	1/3	1/6

$$(C.12)$$

This formula is reminiscent of Simpson's method for numerical integration. The second variant is:

0	0	0	0	0
1/3	1/3	0	0	0
2/3	-1/3	1	0	0
1	1	-1	1	0
	1/8	3/8	3/8	1/8

$$(C.13)$$

This formula is reminiscent of Simpson's 3/8ths rule. Implementation of C.12 is quite simple:

```
void RungKutta4(void dydx(double,double*,double*),
    double*x,double dx,double*y,double*dy,int n)
    {
    int i,j;
    static double a[4]={0.,0.5,0.5,1.};
```

```
static double b[4]={1./6.,1/3.,1./3.,1./6.};
double*w,*v;
w=calloc(4*n,sizeof(double));
v=calloc(  n,sizeof(double));
dydx(x[0],y,w);
for(j=1;j<4;j++)
  {
  for(i=0;i<n;i++)
    {
    dy[i]=a[j]*w[n*(j-1)+i];
    v[i]=y[i]+dx*dy[i];
    }
  dydx(x[0]+dx*a[j],v,w+n*j);
  }
for(i=0;i<n;i++)
  {
  dy[i]=0;
  for(j=0;j<4;j++)
    dy[i]+=b[j]*w[n*j+i];
  y[i]+=dx*dy[i];
  }
x[0]+=dx;
free(w);
free(v);
}
```

Implementation of C.13 just requires changing the preceding a[] and b[] data statements. User-defined function dydx returns the differential:

```
void dydx(double x,double*y,double*dy)
  {
  dy[0]=f1(x,y);
  dy[1]=f2(x,y);
  etc.
  }
```

The stepping process is done inside a loop:

```
for(i=0;i<steps;i++)
  RungKutta4(dydx,&x,dx,y,dy,n);
```

where n is the number of variables.

90

Appendix D. Validation of PTRAX

PTRAX was developed in 1995 by a team at Environmental Consulting Engineers, Inc., under contract to Martin Marietta Energy Systems, Inc., for the U. S. Department of Energy and later for the U. S. Department of Defense. All of the associated documents are available through the Freedom of Information Act (FIOA). The development team consisted of Steve Young, Nick Williams, and myself. Steve was the geohydrologist, Nick built all of the test cases and ran all of the validations, and I wrote the PTRAX software in the C programming language, targeting the Windows® NT operating system. The complete validation report can be downloaded from this link:

https://dudleybenton.altervista.org/publications/Description and Verification of PTRAX.pdf

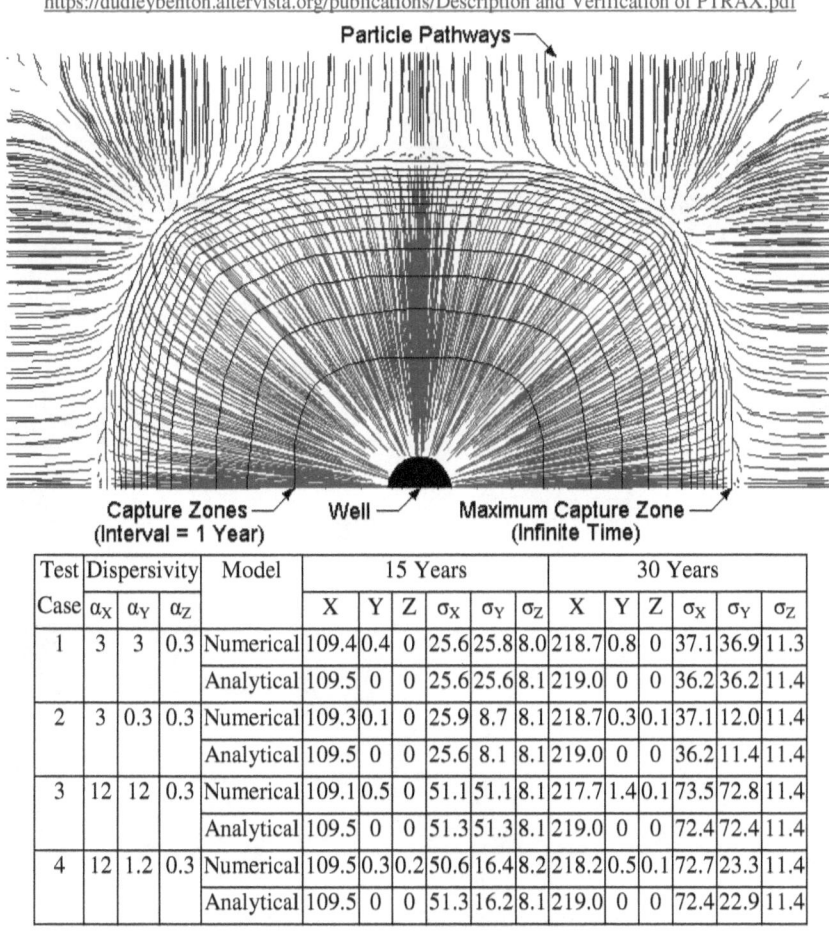

Test	Dispersivity			Model	15 Years						30 Years					
Case	α_X	α_Y	α_Z		X	Y	Z	σ_X	σ_Y	σ_Z	X	Y	Z	σ_X	σ_Y	σ_Z
1	3	3	0.3	Numerical	109.4	0.4	0	25.6	25.8	8.0	218.7	0.8	0	37.1	36.9	11.3
				Analytical	109.5	0	0	25.6	25.6	8.1	219.0	0	0	36.2	36.2	11.4
2	3	0.3	0.3	Numerical	109.3	0.1	0	25.9	8.7	8.1	218.7	0.3	0.1	37.1	12.0	11.4
				Analytical	109.5	0	0	25.6	8.1	8.1	219.0	0	0	36.2	11.4	11.4
3	12	12	0.3	Numerical	109.1	0.5	0	51.1	51.1	8.1	217.7	1.4	0.1	73.5	72.8	11.4
				Analytical	109.5	0	0	51.3	51.3	8.1	219.0	0	0	72.4	72.4	11.4
4	12	1.2	0.3	Numerical	109.5	0.3	0.2	50.6	16.4	8.2	218.2	0.5	0.1	72.7	23.3	11.4
				Analytical	109.5	0	0	51.3	16.2	8.1	219.0	0	0	72.4	22.9	11.4

The original validation consisted of well capture and comparison to an analytical model, AT123D. Both tests were successful. Over the next several

91

years, I added some features and the model was validated with considerable field data with surprisingly good results. A selection of figures is included here. These simulations were performed with 800,000 particles.

Analytical Solution - 15 Years

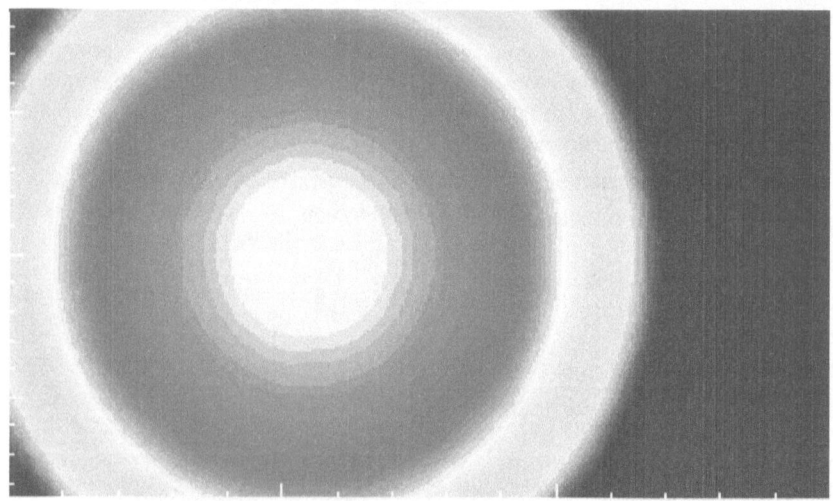

PTRAX - 15 Years

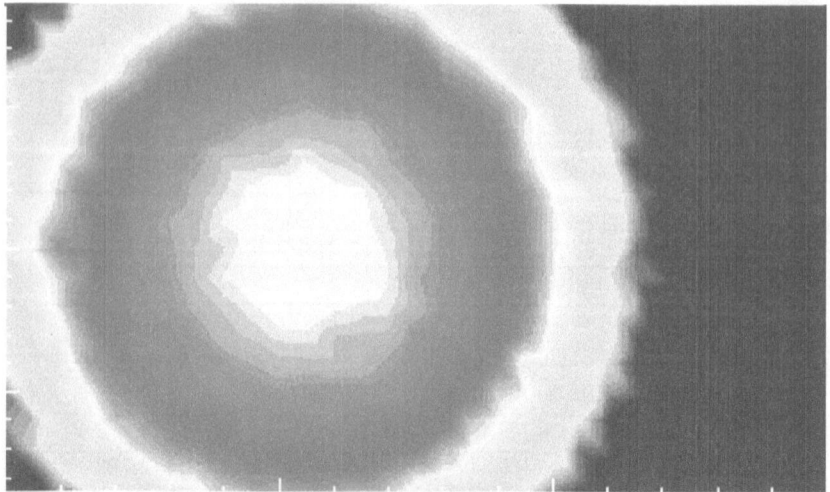

Dispersion Resulting from ax=ay=12m

Analytical Solution - 30 Years

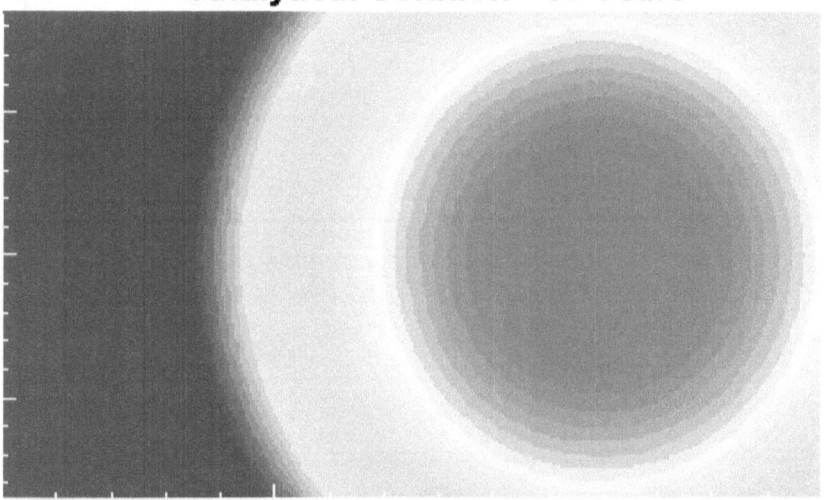

PTRAX - 30 Years

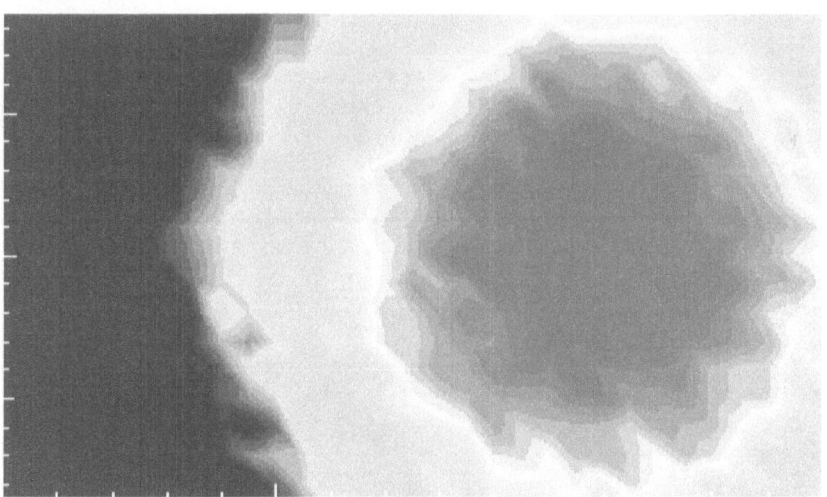

Dispersion Resulting from ax=ay=12m

Analytical Solution - 15 Years

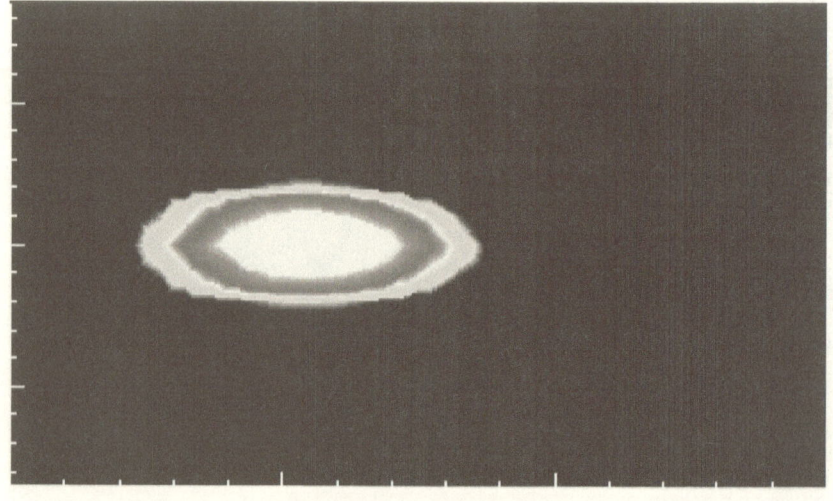

PTRAX - 15 Years

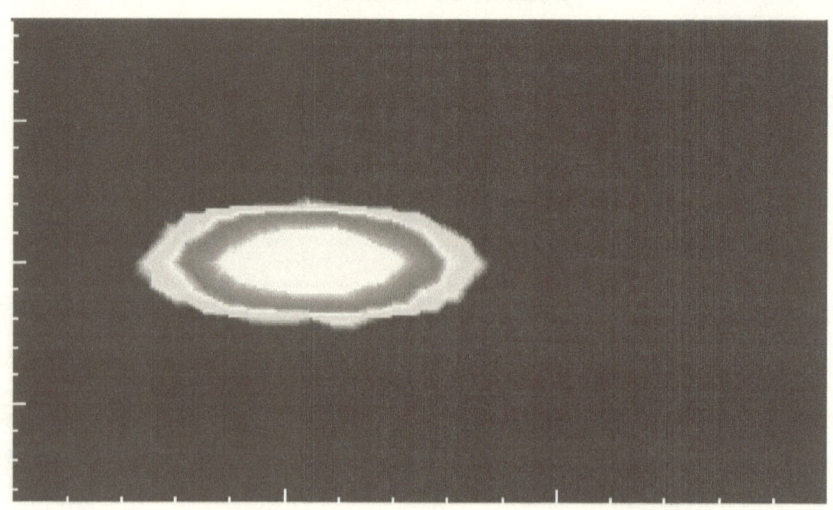

Dispersion Resulting from ax=3m, ay=0.3m

Analytical Solution - 30 Years

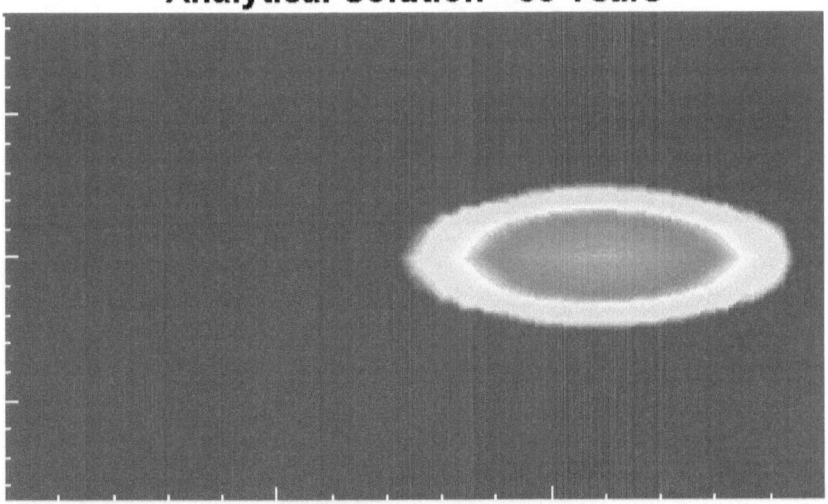

PTRAX - 30 Years

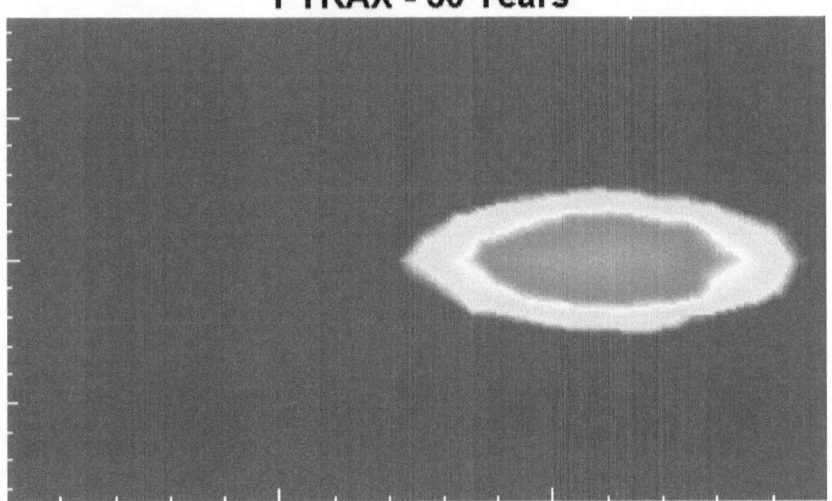

Dispersion Resulting from ax=3m, ay=0.3m

Additional publications are also available, including:

https://dudleybenton.altervista.org/publications/Development of the Fast 3D Particle Tracker PTRAX.pdf

Appendix E. PTRAX Coding

The particle tracking code is written entirely in the C programming language. The code is highly optimized, as evidenced by the illustration in Chapter 15 of tracking three million particles in fourteen minutes on a single 3GHz Intel® processor. There are no elements of object-oriented programming in PTRAX. While such might be efficient for development, object-oriented programming is anything but efficient for execution. This also means that all dynamic arrays are purposely allocated and initialized, not leaving this important task to throwing and catching exceptions. Exceptions (in the object-oriented programming sense) should be reserved for fatal application errors and never used for mundane tasks, like uninitialized variables. Just because the pizza place is next door to the firehouse doesn't mean that pulling the fire alarm and dispatching the big red truck is an efficient way to deliver pizza.

Geometric Functions

One of the most time-consuming aspects of particle tracking through a discrete domain is figuring out which element a particle is in and which ones it might enter next. Below are some of the basic functions used:

```
void EquationLine(double*Xp,double*Yp,double*C) /*
   general equation of a line */
{ /* equation: C0*X+C1*Y=C2 */
C[0]=Yp[1]-Yp[0];
C[1]=Xp[0]-Xp[1];
C[2]=Xp[0]*Yp[1]-Yp[0]*Xp[1];
}
void EquationPlane(double*Xp,double*Yp,
   double*Zp,double*C) /* general equation of a plane */
{ /* equation: C0*X+C1*Y+C2*Z=C3 */
double Y01,Y02,Y12,Z01,Z02,Z12;
Z01=Zp[1]-Zp[0];
Z02=Zp[2]-Zp[0];
Z12=Zp[2]-Zp[1];
Y01=Yp[1]-Yp[0];
Y02=Yp[2]-Yp[0];
Y12=Yp[2]-Yp[1];
C[0]= Yp[0]*Z12-Yp[1]*Z02+Yp[2]*Z01;
C[1]=-Xp[0]*Z12+Xp[1]*Z02-Xp[2]*Z01;
C[2]= Xp[0]*Y12-Xp[1]*Y02+Xp[2]*Y01;
C[3]=C[0]*Xp[0]+C[1]*Yp[0]+C[2]*Zp[0];
}
double Line2Point(double*Xp,double*Yp,double X,double Y)
{
double C[3],H;
EquationLine(Xp,Yp,C);
H=_hypot(C[0],C[1]);
if(H>DBL_EPSILON)
   return((C[0]*X+C[1]*Y-C[2])/H);
```

```
    else
      return(0);
    }
double Point2Line(double*Xp,double*Yp,double X,double Y)
    {
    double A,B,C,S,R;
    A=_hypot(X-Xp[0],Y-Yp[0]);
    B=_hypot(X-Xp[1],Y-Yp[1]);
    C=_hypot(Xp[0]-Xp[1],Yp[0]-Yp[1]);
    S=min(A,B);
    R=max(A,B);
    if(R*R-S*S>C*C)
      return(S);
    else
      return(Line2Point(Xp,Yp,X,Y));
    }
double Plane2Point(double*Xp,double*Yp,double*Zp,double
    X,double Y,double Z)
    {
    double C[4],H;
    EquationPlane(Xp,Yp,Zp,C);
    H=hypot3D(C[0],C[1],C[2]);
    if(H>DBL_EPSILON)
      return((C[0]*X+C[1]*Y+C[2]*Z-C[3])/H);
    else
      return(0);
    }
double Point2Plane(double*Xp,double*Yp,double*Zp,double
    X,double Y,double Z)
    {
    double D1,D2,D3,Dm,Dx,S1,S2,S3,Sm,Sx;
    D1=hypot3D(X-Xp[0],Y-Yp[0],Z-Zp[0]);
    D2=hypot3D(X-Xp[1],Y-Yp[1],Z-Zp[1]);
    D3=hypot3D(X-Xp[2],Y-Yp[2],Z-Zp[2]);
    Dm=min(D1,min(D2,D3));
    Dx=max(D1,max(D2,D3));
    S1=hypot3D(Xp[1]-Xp[0],Yp[1]-Yp[0],Zp[1]-Zp[0]);
    S2=hypot3D(Xp[2]-Xp[1],Yp[2]-Yp[1],Zp[2]-Zp[1]);
    S3=hypot3D(Xp[0]-Xp[2],Yp[0]-Yp[2],Zp[0]-Zp[2]);
    Sm=min(S1,min(S2,S3));
    Sx=max(S1,max(S2,S3));
    if(Dx*Dx-Dm*Dm>Sx*Sx)
      return(Dm);
    else
      return(Plane2Point(Xp,Yp,Zp,X,Y,Z));
    }
double Point2Edge(double*X,double*Y,double*Z)
    {
    double D1,D2,D3;
```

```
D1=hypot3D(X[1]-X[0],Y[1]-Y[0],Z[1]-Z[0]);
if(D1<Ltiny)
   return(D1);
D2=hypot3D(X[2]-X[0],Y[2]-Y[0],Z[2]-Z[0]);
if(D2<Ltiny)
   return(D2);
if((X[1]-X[0])*(X[2]-X[0])>0)
   return(min(D1,D2));
if((Y[1]-Y[0])*(Y[2]-Y[0])>0)
   return(min(D1,D2));
if((Z[1]-Z[0])*(Z[2]-Z[0])>0)
   return(min(D1,D2));
D3=hypot3D(X[2]-X[1],Y[2]-Y[1],Z[2]-Z[1]);
if(D3<Ltiny)
   return(min(D1,D2));
return(2*AreaPlane(X,Y,Z)/D3);
}
double IntersectCircle(double Xp,double Yp,double
   U,double V,double Xw,double Yw,double Rw)
{
double A,B,C,D,D2,dX,dY,P,R2,T; /* compute the closest
   */
/* intersection between      */
A=U*U+V*V; /* a ray and a circle          */
dX=Xp-Xw; /* the ray begins at Xp,Yp  */
dY=Yp-Yw; /* and extends along U,V    */
R2=Rw*Rw; /* the circle has radius Rw */
D2=dX*dX+dY*dY; /* and center Xw,Yw         */
if(D2<=R2) /* if the velocity is zero    */
   return(0); /* then an intersection can    */
else if(A<DBL_EPSILON) /* only occur if the particle
   */
   return(-1); /* is already within the      */
/* capture radius            */
/* for an intersection to occur */
P=V*dX-U*dY; /* the minimum distance between */
if(P*P>A*R2) /* the ray and the well center  */
   return(-1); /* must not be greater than      */
/* the capture radius           */
B=2*(U*dX+V*dY);
C=D2-R2; /* the point of intersection */
D=B*B-4*A*C; /* results in a quadratic     */
if(D<0) /* equation which must have  */
   return(-1); /* real roots (i.e., Dò0)     */
A*=2;
if(D>0) /* avoid as many operations */
   D=sqrt(D)/A; /* as possible, especially û */
B/=A;
T=-B-D;
```

```c
  if(T>=0) /* return the smallest */
    return(T); /* non-negative root   */
  else
    return(-B+D);
  }
double IntersectCylinder(double Xp,double Yp,double
    Zp,double U,double V,double W,double Xw,double
    Yw,double B,double T,double Rw)
  {
  double D2,dX,dY,R2,T1,T2,Tc;
  dX=Xp-Xw;
  dY=Yp-Yw;
  D2=dX*dX+dY*dY;
  R2=Rw*Rw;
  if(D2<=R2) /* test for particle */
    if(B<=Zp&&Zp<=T) /* already in cylinder */
      return(0);
  if(fabs(W)>DBL_EPSILON) /* compute time to intersect
    */
    { /* upper and lower ends */
    T1=(B-Zp)/W;
    T2=(T-Zp)/W; /* if W÷0 then no intersection */
    } /* is possible unless Z lies */
  else if(Zp<B||Zp>T) /* within the cylinder already */
    return(-1);
  /* compute time to */
  Tc=IntersectCircle(Xp,Yp,U,V,Xw,Yw,Rw); /* intersect
    circle */
  if(Tc>=0)
    if(min(T1,T2)<=Tc&&Tc<=max(T1,T2))
      return(Tc);
  return(-1);
  }
```

Animations and Concentrations

Updating the animations and concentration maps as each particle is tracked so as to create the ensemble impact is perhaps the second most time-consuming process.

```c
void UpdateParticles(int why,int steps)
  {
  BYTE b;
  int c,d,m,r,step,t,x,y,z;
  double T,T1,T2,X,Y,Z;
  if((!keep_trap)&&(why==TRAPPED))
    return;
  if(seed_fils<2)
    track_color=brand(blue,red);
  for(t=step=0,T=Tmin;t<Show.Nt;t++,T+=Show.dS)
    {
```

```
if(T<Time[0]) /* skip snapshots before */
   continue; /* particle is seeded     */
if(T>Time[steps-1]) /* test for snapshot later */
   break; /* than particle lifetime  */
while(step<steps-1&&T>Time[step+1]) /* locate
particle time       */
   step++; /* step containing snapshot */
T1=Time[step]; /* time at beginning of step */
T2=max(T1+Ttiny,Time[step+1]); /* time at end of
step */
X=((T2-T)*Xloc[step]+(T-T1)*Xloc[step+1])/(T2-T1);
x=(int)((X-Xmin)/Show.dX);
if(EorF[step]<0&&x>0&&x<Nx-1)
   x+=(rand()%3)-1;
Y=((T2-T)*Yloc[step]+(T-T1)*Yloc[step+1])/(T2-T1);
y=(int)((Y-Ymin)/Show.dY);
if(EorF[step]<0&&y>0&&y<Ny-1)
   y+=(rand()%3)-1;
if(Nd>2)
   {
   Z=((T2-T)*Zloc[step]+(T-T1)*Zloc[step+1])/(T2-T1);
   z=(int)((Z-Zmin)/Show.dZ);
   if(EorF[step]<0&&z>0&&z<Nz-1)
      z+=(rand()%3)-1;
   }
d=(int)((T-Tmin)/Show.dT);
if(file_part&4)
   {
   c=y;
   r=z;
   }
else if(file_part&2)
   {
   c=x;
   r=z;
   }
else
   {
   c=x;
   r=y;
   }
if(c<0||c>=Show.Nc)
   continue;
if(r<0||r>=Show.Nr)
   continue;
if(d<0||d>=Show.Nd)
   continue;
m=(c*Show.Nr+r)*Show.Nd+d;
b=Show.trk[m];
```

```
    if(b>black&&b<white)
      Show.trk[m]=track_color;
    }
  }
void UpdateConcentration(int why,int steps)
  {
  int c,d,m,r,step,x,y,z;
  double M,M1,M2,T,T1,T2,X,Y,Z;
  if((!keep_trap)&&(why==TRAPPED))
    return;
  for(d=step=0,T=Tmin;d<Show.Nd;d++,T+=Show.dT)
    {
    if(T<Time[0]) /* skip snapshots before */
      continue; /* particle is seeded     */
    if(T>Time[steps-1]) /* test for snapshot later */
      break; /* than particle lifetime  */
    while(step<steps-1&&T>Time[step+1]) /* locate
    particle time      */
      step++; /* step containing snapshot */
    T1=Time[step]; /* time at beginning of step */
    T2=max(T1+Ttiny,Time[step+1]); /* time at end of
    step */
    M1=Mass[step];
    M2=Mass[step+1];
    if(M1>0&&M2>0)
      M=M1*Exp(log(M2/M1)*(T-T1)/(T2-T1));
    else
      M=((T2-T)*M1+(T-T1)*M2)/(T2-T1);
    X=((T2-T)*Xloc[step]+(T-T1)*Xloc[step+1])/(T2-T1);
    x=(int)((X-Xmin)/Show.dX);
    if(EorF[step]<0&&x>0&&x<Nx-1)
      x+=(rand()%3)-1;
    Y=((T2-T)*Yloc[step]+(T-T1)*Yloc[step+1])/(T2-T1);
    y=(int)((Y-Ymin)/Show.dY);
    if(EorF[step]<0&&y>0&&y<Ny-1)
      y+=(rand()%3)-1;
    if(Nd>2)
      {
      Z=((T2-T)*Zloc[step]+(T-T1)*Zloc[step+1])/(T2-T1);
      z=(int)((Z-Zmin)/Show.dZ);
      if(EorF[step]<0&&z>0&&z<Nz-1)
        z+=(rand()%3)-1;
      }
    if(file_conc&4)
      {
      c=y;
      r=z;
      }
    else if(file_conc&2)
```

```
        {
        c=x;
        r=z;
        }
    else
        {
        c=x;
        r=y;
        }
    if(c<0||c>=Show.Nc)
        continue;
    if(r<0||r>=Show.Nr)
        continue;
    m=(c*Show.Nr+r)*Show.Nd+d;
    if(Show.con[m]>black&&Show.con[m]<white)
        Conc[m]+=(float)(M/eV);
    }
if(why==TRAPPED||why==CIRCULATION)
    {
    if(d<Show.Nd)
        {
        step=steps-1;
        M=Mass[step];
        x=(int)((Xloc[step]-Xmin)/Show.dX);
        y=(int)((Yloc[step]-Ymin)/Show.dY);
        if(Nd>2)
            z=(int)((Zloc[step]-Zmin)/Show.dZ);
        if(file_conc&4)
            {
            c=y;
            r=z;
            }
        else if(file_conc&2)
            {
            c=x;
            r=z;
            }
        else
            {
            c=x;
            r=y;
            }
        if(c>=0&&c<Show.Nc)
            {
            if(r>=0&&r<Show.Nr)
                {
                m=(c*Show.Nr+r)*Show.Nd+d;
                while(d<Show.Nd)
                    {
```

103

```
            if(Show.con[m]>black&&Show.con[m]<white)
              Conc[m]+=(float)(M/eV);
           d++;
           m++;
            }
         }
       }
     }
   }
 }
```

also by D. James Benton

3D Articulation: Using OpenGL, ISBN-9798596362480, Amazon, 2021 (book 3 in the 3D series).

3D Models in Motion Using OpenGL, ISBN-9798652987701, Amazon, 2020 (book 2 in the 3D series.

3D Rendering in Windows: How to display three-dimensional objects in Windows with and without OpenGL, ISBN-9781520339610, Amazon, 2016 (book 1 in the 3D series).

A Synergy of Short Stories: The whole may be greater than the sum of the parts, ISBN-9781520340319, Amazon, 2016.

Azeotropes: Behavior and Application, ISBN-9798609748997, Amazon, 2020.

bat-Elohim: Book 3 in the Little Star Trilogy, ISBN-9781686148682, Amazon, 2019.

Boilers: Performance and Testing, ISBN: 9798789062517, Amazon 2021.

Combined 3D Rendering Series: 3D Rendering in Windows®, 3D Models in Motion, and 3D Articulation, ISBN-9798484417032, Amazon, 2021.

Complex Variables: Practical Applications, ISBN-9781794250437, Amazon, 2019.

Compression & Encryption: Algorithms & Software, ISBN-9781081008826, Amazon, 2019.

Computational Fluid Dynamics: an Overview of Methods, ISBN-9781672393775, Amazon, 2019.

Computer Simulation of Power Systems: Programming Strategies and Practical Examples, ISBN-9781696218184, Amazon, 2019.

Contaminant Transport: A Numerical Approach, ISBN-9798461733216, Amazon, 2021.

CPUnleashed! Tapping Processor Speed, ISBN-9798421420361, Amazon, 2022.

Curve-Fitting: The Science and Art of Approximation, ISBN-9781520339542, Amazon, 2016.

Death by Tie: It was the best of ties. It was the worst of ties. It's what got him killed., ISBN-9798398745931, Amazon, 2023.

Differential Equations: Numerical Methods for Solving, ISBN-9781983004162, Amazon, 2018.

Equations of State: A Graphical Comparison, ISBN-9798843139520, Amazon, 2022.

Evaporative Cooling: The Science of Beating the Heat, ISBN-9781520913346, Amazon, 2017.

Forecasting: Extrapolation and Projection, ISBN-9798394019494, Amazon 2023.

Heat Engines: Thermodynamics, Cycles, & Performance Curves, ISBN-9798486886836, Amazon, 2021.

Heat Exchangers: Performance Prediction & Evaluation, ISBN-9781973589327, Amazon, 2017.

Heat Recovery Steam Generators: Thermal Design and Testing, ISBN-9781691029365, Amazon, 2019.

Heat Transfer: Heat Exchangers, Heat Recovery Steam Generators, & Cooling Towers, ISBN-9798487417831, Amazon, 2021.

Heat Transfer Examples: Practical Problems Solved, ISBN-9798390610763, Amazon, 2023.

The Kick-Start Murders: Visualize revenge, ISBN-9798759083375, Amazon, 2021.

Jamie2: Innocence is easily lost and cannot be restored, ISBN-9781520339375, Amazon, 2016-18.

Kyle Cooper Mysteries: Kick Start, Monte Carlo, and Waterfront Murders, ISBN-9798829365943, Amazon, 2022.

The Last Seraph: Sequel to Little Star, ISBN-9781726802253, Amazon, 2018.

Little Star: God doesn't do things the way we expect Him to. He's better than that! ISBN-9781520338903, Amazon, 2015-17.

Living Math: Seeing mathematics in every day life (and appreciating it more too), ISBN-9781520336992, Amazon, 2016.

Lost Cause: If only history could be changed..., ISBN-9781521173770, Amazon, 2017.

Mass Transfer: Diffusion & Convection, ISBN-9798702403106, Amazon, 2021.

Mill Town Destiny: The Hand of Providence brought them together to rescue the mill, the town, and each other, ISBN-9781520864679, Amazon, 2017.

Monte Carlo Murders: Who Killed Who and Why, ISBN-9798829341848, Amazon, 2022.

Monte Carlo Simulation: The Art of Random Process Characterization, ISBN-9781980577874, Amazon, 2018.

Nonlinear Equations: Numerical Methods for Solving, ISBN-9781717767318, Amazon, 2018.

Numerical Calculus: Differentiation and Integration, ISBN-9781980680901, Amazon, 2018.

Numerical Methods: Nonlinear Equations, Numerical Calculus, & Differential Equations, ISBN-9798486246845, Amazon, 2021.

Orthogonal Functions: The Many Uses of, ISBN-9781719876162, Amazon, 2018.

Overwhelming Evidence: A Pilgrimage, ISBN-9798515642211, Amazon, 2021.

Plumes: Delineation & Transport, ISBN-9781702292771, Amazon, 2019.

Power Plant Performance Curves: for Testing and Dispatch, ISBN-9798640192698, Amazon, 2020.

Practical Linear Algebra: Principles & Software, ISBN-9798860910584, Amazon, 2023.

Props, Fans, & Pumps: Design & Performance, ISBN-9798645391195, Amazon, 2020.

Remediation: Contaminant Transport, Particle Tracking, & Plumes, ISBN-9798485651190, Amazon, 2021.

ROFL: Rolling on the Floor Laughing, ISBN-9781973300007, Amazon, 2017.

Seminole Rain: You don't choose destiny. It chooses you, ISBN-9798668502196, Amazon, 2020.

Septillionth: 1 in 10^{24}, ISBN-9798410762472, Amazon, 2022.

Software Development: Targeted Applications, ISBN-9798850653989, Amazon, 2023.

Software Recipes: Proven Tools, ISBN-9798815229556, Amazon, 2022.

Steam 2020: to 150 GPa and 6000 K, ISBN-9798634643830, Amazon, 2020.

Thermochemical Reactions: Numerical Solutions, ISBN-9781073417872, Amazon, 2019.

Thermodynamic and Transport Properties of Fluids, ISBN-9781092120845, Amazon, 2019.

Thermodynamic Cycles: Effective Modeling Strategies for Software Development, ISBN-9781070934372, Amazon, 2019.

Thermodynamics - Theory & Practice: The science of energy and power, ISBN-9781520339795, Amazon, 2016.

Version-Independent Programming: Code Development Guidelines for the Windows® Operating System, ISBN-9781520339146, Amazon, 2016.

The Waterfront Murders: As you sow, so shall you reap, ISBN-9798611314500, Amazon, 2020.

Weather Data: Where To Get It and How To Process It, ISBN-9798868037894, Amazon, 2023.